# Evolutionary Pathways in Nature

## A Phylogenetic Approach

Reconstructing phylogenetic trees from DNA sequences has become a popular exercise in many branches of biology, and here the award-winning geneticist John Avise explains why. Molecular phylogenies provide a genealogical backdrop for interpreting the evolutionary histories of many other types of biological traits (anatomical, behavioral, ecological, physiological, biochemical, and even geographical). Guiding readers on a natural history tour along dozens of evolutionary pathways, the author describes how creatures ranging from microbes to elephants came to possess their current phenotypes. If you want to know how the toucan got its bill and the kangaroo its hop, then this is the book for you. This book also provides a definitive answer to the proverbial question: 'which came first, the chicken or the egg?' This scientifically educational yet entertaining treatment of ecology, genetics and evolution is intended for college students, professional biologists, and anyone interested in natural history and biodiversity.

**John C. Avise** is a distinguished professor of Ecology and Evolutionary Biology at the University of California, Irvine.

# Evolutionary Pathways in Nature

## A Phylogenetic Approach

**JOHN C. AVISE**

Department of Ecology and Evolutionary Biology
University of California

Illustrations by Trudy Nicholson

CAMBRIDGE UNIVERSITY PRESS
Cambridge, New York, Melbourne, Madrid, Cape Town, Singapore, São Paulo

Cambridge University Press
The Edinburgh Building, Cambridge CB2 2RU, UK

Published in the United States of America by Cambridge University Press, New York

www.cambridge.org
Information on this title: www.cambridge.org/9780521857536

© John C. Avise 2006

First published 2006

Printed in the United Kingdom at the University Press, Cambridge

*A catalog record for this publication is available from the British Library*

ISBN-13  978-0-521-85753-6 hardback
ISBN-10  0-521-85753-8 hardback

ISBN-13  978-0-521-67417-1 paperback
ISBN-10  0-521-67417-4 paperback

# CONTENTS

Many biologists now incorporate molecular phylogenetic analyses into their explorations of nature. Using sophisticated laboratory techniques, they uncover "DNA markers" or "genetic tags" that uniquely identify each creature. Furthermore, details in the submicroscopic structures of these natural labels offer tantalizing clues to how living organisms were genealogically linked through bygone ancestors. Thus, lengthy DNA sequences housed in the cells of all organisms carry not only the necessary molecular genetic instructions for life, but also extensive records of phylogeny, i.e. of evolutionary ancestry and descent.

During the replication and transmission of DNA from one generation to the next, mutations continually arise. Many of these spread through populations (via natural selection, or sometimes by chance genetic drift), thereby cumulatively altering particular molecular passages in each species' hereditary script. In recent years, scientists have learned how to read and interpret the genealogical content of these evolutionary diaries – these "genomic autobiographies" – of nature. Results are summarized as phylogenetic diagrams that depict how particular forms of extant life are connected to one another via various historical branches in the Tree of Life.

Phylogenetic analysis has become a wildly popular exercise in many areas of biology, but phylogenies estimated from DNA sequences are seldom the ultimate objects of scientific interest. The primary value of each molecular phylogeny lies instead in its utility as historical backdrop for deciphering the evolutionary histories of other kinds of biological traits such as morphologies, physiologies, behaviors, lifestyles, or geographical distributions. By mapping such organismal features onto species' phylogenies estimated from molecular data, biologists can address fascinating questions of the following sort. Did the bipedal hop arise once or multiple times in kangaroo evolution? From what type of ancestor did toucan birds evolve their banana-like bills? How often during evolution have reptiles lost their limbs? Are the antifreeze proteins in Arctic and Antarctic fishes functionally similar by virtue of shared ancestry or convergent evolution? By what evolutionary routes have some fishes evolved powerful electrical discharges? Did Jamaican land crabs derive their peculiar forms of offspring care from a common ancestor? Did walkingstick insects evolve from flyingsticks or vice versa, and how often? How have certain bacteria acquired their magnetic compasses? On how many occasions have distinct algal and fungal lineages joined forces in lichen symbioses? Where on the planet have phylogenetic appraisals uncovered cryptic species and conservation-relevant hotspots of global biodiversity? Can the ancient breakup of the supercontinent Gondwanaland account for the modern distributions of particular lineages of birds and mammals in the Southern Hemisphere? Where and when did the viruses responsible for the AIDS epidemic enter the human species? And, which came first: the chicken or the egg?

By highlighting studies that have provided scientific answers to these and many additional questions, I intend to illustrate the power (and also some limitations) of comparative phylogenetic perspectives in biological research. Several available textbooks describe, in depth, how molecular data are gathered in the laboratory and analyzed at the computer. My approach here will not be to recount the many operational details of molecular phylogenetics (although introductory background is provided). Rather, my intent is to serve as a naturalist guide on a biological expedition into the remarkable world of nature, as viewed through the evolutionary prism of molecular phylogeny. In each of 67 essays arranged into six topical chapters, I describe how a DNA-estimated phylogeny provided historical framework for interpreting a puzzling ecological feature or evolutionary process in organisms with unusual anatomies or lifestyles, or in creatures with special significance to one or another biological field such as ethology, natural history, biogeography, conservation, biochemistry, physiology, epidemiology, or medicine.

Through this case-history approach, I hope to provide a fun yet educational introduction – for amateur naturalists and students to professional biologists – to how comparative phylogenetic analyses have helped to solve some of nature's most intriguing mysteries. Another goal is to encourage a deeper appreciation of the many intellectual and aesthetic treasures of the biological world. As more and more people become educated about nature's ways, perhaps societies will learn to cherish life's variety and strive harder to preserve what remains. Tragically, through human actions, populations and species today are being driven to extinction at rates seldom experienced in the planet's long history. To terminate any lineage now is to lose forever a genetic wisdom that was honed along an epic evolutionary journey lasting nearly four billion years. Paradoxically, life is both fragile and tenacious. Extinction continually threatens, and once realized can never be undone. However, having withstood and adapted to countless environmental challenges over the geological eons, each extant lineage is also a hardy and proven survivor, surely deserving of our deepest respect and admiration.

## ACKNOWLEDGMENTS

Doug Futuyma, Blair Hedges, David Hillis, Kirk Jensen, Judith Mank, Axel Meyer, David Reznick, DeEtte Walker, John Ware, and several anonymous reviewers provided helpful comments on various portions of the text. I am especially grateful to Trudy Nicholson for producing the beautiful plant and animal drawings that grace this book.

# 1 | Introduction

Long before the concept of biological evolution entered the human mind, people classified diverse forms of life into recognizable categories. Some of the earliest spoken words undoubtedly were names ascribed by primitive peoples to particular types of plants and animals important in their daily lives. Theorists and professional biologists categorized organisms too. For example, in the third century BC the Greek philosopher Aristotle grouped species according to morphological conditions (such as winged versus wingless, and two-legged versus four-legged) that he supposed had been constant since the time of Creation. About twenty centuries later, Carolus Linnaeus – a Swedish botanist and the acknowledged father of biological taxonomy – classified organisms into nested groups (such as genera within families within orders within classes), but still he had no inkling that varied depths of evolutionary kinship might underlie these hierarchical resemblances.

More time would pass before scientists finally began to understand that life evolves, and that historical descent from shared ancestors was responsible for many of the morphological similarities among living (and fossil) species. This epiphany is sometimes mistakenly attributed to Charles Darwin (CD), but several scientists before him in the late 1700s and early 1800s, including Jean-Baptiste Lamarck, Comte de Buffon, and CD's own grandfather Erasmus Darwin, were well aware of the reality of evolutionary descent with modification. What CD "merely" added was the elucidation of natural selection as the primary driving agent of adaptive evolution (this achievement was, of course, one of the most influential in the history of science). The point here, however, is that even before CD, the nested classifications of traditional taxonomy had been interpreted by some systematists as logical reflections of the nested branching structures in evolutionary trees.

## The meaning of phylogeny

Evolution has few universal laws, but one unassailable truth is that every organism alive today had at least one parent, who in turn had either one or two parents (depending on whether the lineage was asexual or sexual), and so on extending back in time. The following imagery may help to convey the incredible temporal durations of these extended hereditary lines. Imagine yourself as the current carrier of a genetic baton that was passed along a multi-generation relay team composed of your direct-line ancestors across the past 200 000 years (*c.* 10 000 generations), beginning when creatures virtually indistinguishable from modern *Homo sapiens* first strolled onto the evolutionary scene. If each successive generation of your predecessors had jogged a quarter mile, your family's cross-country relay squad could have transferred the baton from Los Angeles to New York.

The proto-human lineage is known to have separated from the proto-chimpanzee lineage about five to seven million years ago, so across that longer stretch of geological time your ancestral relay team would have covered a distance equivalent to three times the earth's circumference, or about one-third of the way to the moon. If the evolutionary marathon had been monitored across 40 million years, starting when anthropoid primates first arose, at least two million generations of your ancestors would have come and gone (actually more than that, because monkeys have shorter generation lengths than humans). During that time, your hereditary baton would have been passed a distance of at least half a million miles. This logic can be extended (Dawkins, 2004), ultimately to life's origins nearly four billion years ago. If your extended family lineage had dropped the genetic baton (failed to reproduce) even once during this Olympian marathon, you would not be here. Comparable statements apply to every living creature, each of which is the current embodiment of its own hereditary legacy ultimately stretching back through an *unbroken* chain of descent, with genetic modification, across untold generations.

The word phylogeny (from Greek roots "phyl" meaning tribe or kind, and "geny" meaning origin) refers to the chronicle of life, i.e. to the extended hereditary connections between ancestors and their descendents. Thus, phylogeny can be broadly defined as the evolutionary genetic history of life at any and all temporal scales, ranging from close kinship within and among closely related species to ancient connections between distantly related organisms that last shared common ancestors hundreds of millions of years ago.

In the first 100 years following Darwin, scientists estimated phylogenies for particular taxa by comparing visible organismal phenotypes – e.g. morphological, physiological, or behavioral characteristics – that they could only presume were reflective of underlying genetic relationships. When species were found to share particular phenotypic traits, the usual supposition was that they did so by virtue of shared ancestry. This interpretation was not always correct, however, because some phenotypes arise by convergent evolution. Wings, for example, originated independently in insects, birds, and mammals (plus some other groups, such as the pterodactyl reptiles of the Mesozoic Era). So, among extant vertebrates (animals with backbones) alone, wing-powered flight evolved at least once in birds and again independently in bats, but this fact becomes apparent only when many other phenotypic features (such as feathers, fur, and pregnancy) are taken into account. Few evolutionary cases are so straightforward, however, and the basic challenge in genuinely intriguing situations is to distinguish phenotypic conditions that provide a valid phylogenetic signal from those that yield mostly phylogenetic noise (i.e. homoplasy).

Following the introduction of various molecular technologies, beginning about 40 years ago, scientists gained powerful new genetic tools to estimate phylogenetic trees for any living species, as connected across any depths in the vast contin-uum of evolutionary time. This temporal scope is made possible because some DNA sequences have evolved very rapidly, others very slowly, and others at inter-mediate rates. Fast-evolving sequences are most useful for estimating phylogeny at shallow evolutionary timescales (i.e. for organisms that shared common ancestors within the past few thousands or millions of years), whereas slow-evolving DNA sequences find special utility in estimating phylogeny over much deeper evolu-tionary timeframes. Few types of molecular trait are themselves free of homoplasy, but when hundreds, thousands, or millions of molecular characters are examined in a given study (as is now routinely the case), empirical experience has indicated that they collectively beam a strong phylogenetic signal.

In 1973, the famous evolutionary geneticist Theodosius Dobzhansky encapsu-lated a fundamental biological truth in one pithy statement: "Nothing in biology makes sense except in the light of evolution." It is equally true, in turn, that much in evolution makes even more sense in the light of phylogeny. Biological entities are unlike inorganic units (such as gas molecules, or rocks) that can move rather freely in any direction and speed in response to external forces. Instead, the history-laden genetic makeup of organisms directs and constrains each species to a small subset of all imaginable evolutionary trajectories. Each extant species is a current incarnation of an extended lineage whose idiosyncratic genetic past has dictated the present and will also delimit that species' evolutionary scope for the future. Gorillas may dream of flying, but their ponderous bodies of primate ancestry pre-clude self-powered flight from their foreseeable evolutionary prospects.

## Phylogenetic metaphors

Various metaphors can help to capture the general notion of phylogeny. A for-merly popular (but invalid) metaphor portrayed evolution as a ladder, the rungs of which held successive forms of life that presumably had climbed higher and higher toward biological perfection. The lowest rungs were occupied by "lowly" microbes, and atop the highest rung was, of course, *Homo sapiens*. A metaphor with greater legitimacy describes biological lineages as genetic threads stretching back through the ages, and from which the fabric of life has been woven by natural selection and other evolutionary forces including mutation, recombination, and serendipity. As mentioned above, an ineluctable truth is that any lineage alive today extends back generation after generation, ultimately across several billion years to when life originated. Only a minuscule fraction of such hereditary lineages has persisted across the eons to the present; extinction has been the fate of all others. Quite literally, lineages fortunate enough to have survived this epic evolutionary journey have hung on by just a thread.

The eminent paleontologist George Gaylord Simpson invoked another powerful metaphor when he proclaimed: "The stream of heredity makes phylogeny; in a sense, it is phylogeny. Complete genetic analysis would provide the most priceless data for the mapping of this stream." That statement, issued in 1945, was all the more prescient because it came in the "pre-molecular" era, before direct biochemical assays of DNA were available (indeed, even before DNA was firmly documented to be life's hereditary material). Like other biologists of his time, Simpson estimated phylogenies by comparing morphological features among living and fossil species. He nonetheless appreciated that morphological resemblance is merely a surrogate (and sometimes a rather poor one) for establishing propinquity of descent among the creatures being compared, and that direct genetic analyses eventually would be required. Today, by extracting and comparing DNA sequences from living creatures (and occasionally from well-preserved recent fossils), and by reconstructing phylogenies from those molecular genetic data, scientists can more fully explore both the headwaters and the many forks in the streams that constitute the evolutionary watersheds of life.

Ever since the mid 1800s, however, the most popular metaphor for evolution's pathways has not been ladders, threads, or watersheds, but rather phylogenetic trees (Box 1.1; Fig. 1.1). Under this view, DNA is the sap of heredity that has flowed through the ancient roots, trunks, and branches, and finally into the most recent twigs in various sections of the Tree of Life. The tree analogy for phylogenies is indeed apt (albeit imperfect). Much as twigs and limbs in a botanical tree trace back to successively older forks, so too do living species trace their ancestries back through branched hierarchies of ever-more-ancient phylogenetic nodes. Just as forks in a botanical tree tend to be bifurcate (rather than multi-furcate), most speciational nodes in a phylogenetic tree are dichotomous. Much as a real tree fosters new growth primarily from its growing tips and buds, biodiversity at any point in evolutionary time propagates exclusively from then-extant species.

---

## Box 1.1  Basic definitions regarding phylogenetic trees

See Fig. 1.1 for examples; see also Box A1 in the Appendix, and the Glossary, for additional relevant terms and concepts.

(a) *phylogenetic tree (phylogeny)*: a graphical representation of evolutionary genetic history.

(b) *phylogenetic network*: an unrooted phylogeny (e.g. diagram I in Fig. 1.1).

(c) *root*: the most basal branch (pre-dating the earliest node) in a phylogenetic tree (the thick line at the left in diagrams II and III of Fig. 1.1).

(d) *branch*: an extended ancestral–descendent lineage between nodes in a phylogenetic tree.

(e) *interior node*: a branching point inside a phylogenetic tree (i.e. an ancestral point from which two or more branches stem, or, from the perspective of the present, an ancestral point to which any specified set of extant lineages coalesces). In Fig. 1.1, interior nodes are indicated by black dots labeled with the lower-case letters g–k. In any phylogenetic network, interior nodes can be thought of as ball-and-socket joints around which branches can be freely rotated without materially affecting network structure. Thus, angles between branches have no meaning. Similarly, branches can be rotated around interior nodes in a rooted phylogenetic tree, but only in the vertical plane.

(f) *exterior node*: an outer tip on a phylogenetic network or tree, usually representing an extant species (e.g. A–F in the diagrams of Fig. 1.1).

(g) *operational taxonomic units (OTUs)*: the biological entities (e.g. DNA sequences, individuals, populations, species, or higher taxa) analyzed and depicted in a particular phylogenetic tree (again, A–F in Fig. 1.1).

(h) *anagenesis*: genetic change within a lineage (along one branch of a phylogenetic tree) through time.

(i) *cladogenesis*: the splitting or bifurcation of branches in a phylogenetic tree (normally equated with speciation).

(j) *cladogram*: a representation of cladistic relationships, i.e. of a phylogenetic tree's hierarchical branching structure (but otherwise implying nothing about branch lengths).

(k) *phylogram*: a representation of a phylogenetic tree that includes information on branch lengths in addition to cladistic (branching) relationships.

(l) *phenogram*: a tree-like depiction that summarizes overall phenetic (not necessarily phylogenetic) relationships among a set of organisms.

(m) *gene tree*: a graphical representation of the evolutionary history of a particular genetic locus (as opposed to the composite organismal phylogeny of which any gene tree is only a small component).

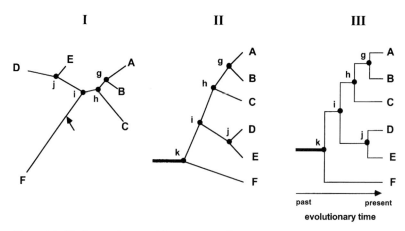

*Figure 1.1.* Phylogenetic trees. Diagrams I–III illustrate alternative but essentially equivalent representations of evolutionary relationships among six hypothetical extant species (A–F). Diagram I is an unrooted tree (phylogenetic network), whereas phylogenetic trees II and III are rooted (at the position of the arrow in diagram I). See Box 1.1 for additional definitions and descriptions.

One shortcoming of the tree metaphor, however, is that real trees have a large trunk and successively smaller branches and twigs, whereas hereditary routes in phylogenetic trees have no distinct tendency to decrease (or increase) in diameter across evolutionary time. In a phylogenetic tree, what split at each node are particular biological species, rather than composite collections of independent species. For example, birds did not evolve from reptiles collectively; rather, one or a few related reptilian species in the Mesozoic Era gave rise to particular proto-avian species from which all other birds eventually descended. For this reason, all phylogenetic trees depicted in this book will be drawn as stick-like diagrams with more or less uniform branch width. In addition, to make labeling easier, nearly all phylogenetic trees presented here will be rotated through 90° relative to an upright real tree, such that the right terminus of each diagram indicates the present time and successive nodes to the left reflect progressively older dates in the evolutionary past.

Charles Darwin included only one figure in his 1859 masterpiece *The Origin of Species*. It was of a phylogenetic tree (albeit an unattractive rendition). However, Ernst Haeckel (a German philosopher and evolutionary biologist) did far more to make an iconography of the tree metaphor by gracing his 1866 book – *Generelle Morphologie der Organismen* – with lovely arbor diagrams, one of which is shown here in Fig. 1.2. Haeckel drew his trees as literal metaphors, complete with bark and gnarled branches. There is, however, a serious shortcoming (apart from the branch-width issue mentioned above) in the style of Haeckel's depictions: namely,

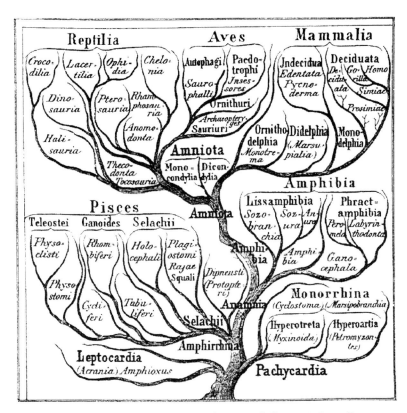

*Figure 1.2.* Example of a phylogenetic tree from Haeckel's (1866) *Generelle Morphologie der Organismen.*

they convey an impression that some living species (such as birds and mammals) are higher in the Tree of Life than others (such as fishes and amphibians), when in fact all lineages leading to extant forms of life have maintained continuous genetic ancestries that trace back ultimately to life's geneses. Thus, if height above the ground in Haeckel's trees is taken to imply the duration of evolutionary existence, the depictions are misleading, because by this criterion all extant branch tips are, in truth, equally high. This is another reason why most of the phylogenetic trees depicted in this book have right-justified branch tips.

Most of our scientific understandings about biology can be improved by implicit or explicit reference to well-grounded phylogenetic trees. For example, the basic challenge in the science of systematics is to describe various portions in the Tree of Life, i.e. to reconstruct the temporal order of forks (speciation events), to measure branch lengths (the amount of genetic change along each branch through time), and to estimate how many buds (distinct species and populations) currently exist from which any future growth might ensue. A primary aim of the conservation sciences is to promote the survival and the potential for diversification of the outermost tips in the Tree of Life. This task is daunting because a burgeoning human population, through its direct and indirect impacts on the environment, threatens to prune if not defoliate much of the Tree's luxuriant canopy. Societies must find better ways to identify, characterize, and protect the vigorous as well as the most tender of extant shoots so that, in this latest instant of geological time, humankind does not terminate what nature had germinated and assiduously propagated across the eons. And finally, in the sciences of ecology, paleontology, ethology, natural history, and evolutionary biology, a fundamental challenge is to understand the historical origins of species and their diverse phenotypes. As I attest via this book, all of these tasks demand an appreciation of phylogeny.

## Molecular appraisals of phylogeny

In 1963, a biochemist in Chicago reported a discovery that would prove to be a major conceptual step forward in the field of phylogenetic biology. By compiling and scrutinizing published information on cytochrome $c$ – a protein involved in cellular energy metabolism, and consisting of a molecular string of 104 amino acids (the building blocks of proteins) – Emanuel Margoliash (1963) found that these molecules differed in structure, to varying degrees, among human, pig, horse, rabbit, chicken, tuna, and baker's yeast. For example, the cytochromes $c$ of horse and pig differed at three amino acid positions along the molecular string, whereas those of horse and tuna differed at 19 amino acid positions and horse and yeast differed at 44 such sites. These differences in amino acid sequence reflect the evolutionary accumulation of underlying mutations in the DNA molecules (i.e. the genes) that encode cytochrome $c$. Margoliash concluded that "The extent of variation among cytochromes $c$ is compatible with the known phylogenetic relations of species. Relatively closely related species show few differences . . . phylogenetically distant species exhibit wider dissimilarities."

From a phylogenetic perspective, there is nothing special about cytochrome $c$. It is merely one of many thousands of cellular proteins, each encoded by a different functional gene. The genes themselves consist of long strings of four types of molecular subunits – the nucleotides adenine (A), thymine (T), cytosine (C), and guanine (G) – that make up not only protein-coding DNA sequences but also vast stretches of non-coding DNA. The collective lengths of these nucleotide strings are astounding. For example, each copy of the human genome (a full suite of DNA in each of our cells) consists of more than three billion pairs of nucleotides wedded into strands that give DNA its double helical structure. The genomes of most other vertebrates are roughly similar in size, and those of various species of invertebrate animals, fungi, and plants range in length from about 10 million to more than 200 billion nucleotide pairs.

Margoliash's findings provided one of the first clear indications that DNA sequences sampled from organismal genomes gradually accumulate specifiable molecular differences during the course of evolution, and that "the extent of variation of the primary structure . . . may give rough approximations of the time elapsed since the lines of evolution leading to any two species diverged." We now know that, during their passage across large numbers of successive generations, DNA molecules (and hence any protein molecules they may encode) often tend to evolve in clock-like fashion. Although molecular clocks are far from metronomic – they tend to tick at somewhat different rates depending on the lineage and on the specific type of DNA sequence examined (see Box 1.2) – they none the less can be informative about approximate nodal dates in evolutionary trees. Furthermore, some methods for estimating phylogenetic trees depend hardly at all on a clock-like behavior (see the Appendix). For example, by considering the evolutionary chains of mutational events required to convert one DNA sequence into another, branching topologies of evolutionary trees can often be recovered even when precise evolutionary dates cannot be attached to particular internal nodes. The bottom line is that, when researchers sample and compare long molecular passages from organisms' genomic archives, they can deduce how various living species have been connected to one another in their near and distant evolutionary pasts.

---

### Box 1.2  DNA sequences for molecular phylogenetics

Many different types of DNA sequence are employed to estimate organismal phylogeny, the choice in each instance dictated by the general evolutionary timeframe under investigation and also by numerous technical considerations. The following are introductory comments about some of the gene sequences widely used in comparative phylogenetics.

**Cytoplasmic genomes**
These are relatively small suites of DNA housed inside organelles within the cytoplasm of eukaryotes (organisms whose cells have a distinct membrane-bound nucleus). The two primary cytoplasmic genomes are mtDNA in the mitochondria of animals and plants, and cpDNA in the chloroplasts of plants.

In animals, mtDNA is usually a closed-circular molecule about 16 000 to 20 000 nucleotide pairs long. It typically consists of 37 functional genes: two ribosomal (r) RNA loci, 22 transfer (t) RNA loci, and 13 structural genes specifying polypeptides (protein subunits) involved in cellular energy production.

The molecule tends to evolve quite rapidly overall, thus making it suitable for phylogenetic appraisals at micro-evolutionary scales (e.g. of conspecific populations), and also across meso-evolutionary timeframes (i.e. for species that separated up to scores of millions of years ago). Different mtDNA loci evolve at quite different rates, however, with some (such as the control region) diverging very rapidly and others (such as the rRNA loci) evolving far more slowly. Thus, with appropriate choice of mtDNA sequences, phylogenetic studies can be tailored to varying evolutionary timescales.

Full-length mtDNA molecules in plants are much larger (200 000 to more than 2 000 000 nucleotide pairs long, depending on the species) and for various technical reasons have not proved particularly useful for phylogenetic reconstructions. By contrast, plant cpDNAs offer powerful phylogenetic markers. These closed-circular molecules, ranging from about 120 000 to 220 000 nucleotide pairs long, generally evolve at a leisurely pace, so their sequences tend to be especially suitable for phylogenetic estimates among plant genera, families, and taxonomic orders.

**Nuclear genomes**
In any eukaryotic cell, most of the tremendous variety of DNA sequences is housed within the nucleus. For example, each complete set (i.e. haploid copy) of the human nuclear genome consists of more than three billion nucleotide pairs arranged along 23 chromosomes. The nuclear genome of a typical eukaryotic species (humans included) contains about 25 000 protein-coding genes, one or a handful of which are normally sequenced from multiple species in a conventional molecular phylogenetic analysis. Useful phylogenetic information can also be recovered from other nuclear regions such as rRNA loci, regulatory domains flanking structural genes, or particular subsets of non-coding sequences (often highly repetitive) that actually make up the great majority of nuclear DNA in most species.

**Combined information**
Typical molecular phylogenetic analyses conducted to date have involved DNA sequences from several nuclear or cytoplasmic genes (or both) totaling about one thousand to several thousand nucleotide pairs per specimen. With continuing improvements in DNA sequencing technologies, the standards are quickly being raised. For example, it has become almost routine in recent years for phylogeneticists to sequence entire 16 kilobase mtDNA genomes from the animals they survey.

Representative genomes of approximately 1000 species (humans included) have been fully sequenced in recent years, and substantial amounts of DNA sequence data are rapidly accumulating for many thousands more. Today, scientists routinely read these genetic scriptures to reconstruct the histories of life. As Margoliash correctly presaged in 1963, molecular details in the genomic registries provide "a faithful recorder of the unit events of evolution." Scientists are no longer content, however, to draw crude phylogenetic sketches for a miscellany of distant organisms such as human, rabbit, chicken, tuna, and yeast. Instead, they now use extensive molecular data to paint detailed evolutionary pictures for hundreds of species of mammals, birds, reptiles, amphibians, and fishes, plus all sorts of invertebrate animals, fungi, plants, and microbes. In the past decade or two, molecular phylogenetics has grown into one of the most active areas in all of biological research. Perhaps 10 million or more species currently inhabit the planet, so reconstruction of the full Tree of Life will require a huge scientific effort. None the less, DNA sequences recovered to date already permit solid phylogenetic estimates for many taxa, and the molecular phylogenies in turn can be employed to chart the evolutionary courses of organismal phenotypes (morphological, behavioral, and so on). This book will delve into some of the most evocative and sometimes controversial of these phylogenetic mapping exercises to date.

## Comparative phylogenetics

Evolutionary biologists routinely employ "comparative phylogenetic" methods, by which can be meant many things. In a catholic sense, any phylogenetic procedure is comparative if it involves more than one gene, more than one phenotype, more than one taxonomic group, or any combination of the above. For example, it is perfectly permissible and often highly informative to compare a phylogenetic estimate based on one set of phenotypes with that based on another set of phenotypes, or to compare the phylogenies of two or more taxonomic groups against a geographic backdrop (for example) that may have influenced their evolutionary histories. In other words, the basic idea of comparative phylogenetics is to compare and contrast historical evolutionary patterns across multiple types of characters or taxa.

The molecular revolution in evolutionary biology, which began in the second half of the twentieth century, ushered in another powerful way to conduct comparative phylogenetics. Specifically, it afforded access to a potentially huge set of DNA-level and protein-level characters that could be employed as a basis for phylogenetic comparisons with organism-level traits. This book will focus on how molecular estimates of phylogeny have informed our understanding of the ways and means by which organismal phenotypes evolve. However, I hope not to be interpreted as chauvinistic with regard to molecular approaches. The fact is that systematists practiced comparative phylogenetics, widely and fruitfully, long before molecular genetic techniques became available. The comparisons then were based on visible phenotypes and other traditional systematic characters, many of which are readily accessible and hugely informative in their own right about phylogenetic relationships. Molecular data have added another comparative layer to phylogenetic practices, thereby enriching a field that already had a long and productive scientific tradition.

As applied here in a narrower sense, the term comparative phylogenetics will mean any application of DNA-based phylogenies with the intent of revealing the evolutionary histories of organismal phenotypes. In this explicit subcategory of comparative phylogenetic analysis, four steps are normally entailed: (i) DNA-sequencing methods or other laboratory techniques are used to gather extensive molecular data from homologous genes in living species; (ii) based on that genetic data, a phylogeny for those species is estimated by using appropriate tree-building algorithms; (iii) particular phenotypic characters showing variation (such as winged versus wingless) among the species of interest are examined to establish their present-day taxonomic distributions; and (iv) the phylogenetic histories of those phenotypes are provisionally reconstructed by plotting their inferred ancestral states and evolutionary interconversions along various branches of a molecular tree. The first three steps can be thought of as background and the fourth step as the crux of the process of "phylogenetic character mapping" (see below).

A vast technical literature exists on molecular methodologies and phylogenetic algorithms underlying steps (i) and (ii). A cursory introduction is provided in Box 1.3, but readers interested in further details are directed elsewhere (see the references listed). Fortunately, for current purposes, a thorough understanding of molecular techniques and phylogenetic reconstruction methods is not a prerequisite for appreciating the biological discoveries about nature that are the focal points of this book.

Step (iii) normally involves relatively straightforward description, except that questions may arise about how to define and characterize organismal phenotypes. For example, these phenotypes may be alternative qualitative states of composite structures (such as wing-presence versus wing-absence) or behaviors (e.g. flighted versus flightless), or more narrowly defined characters (such as wings made of skin flaps versus those made of feathers and with interior bones; or flapping versus gliding flight). Always, evolutionary interpretations must be adjusted accordingly. For example, the broad attribute "flight" is clearly polyphyletic (i.e. arose on multiple evolutionary occasions) in animals, whereas more specific characteristics often associated with flight (such as feathers in birds, echolocation in bats, or presence

## Box 1.3  Molecular methods and phylogenetic algorithms

Steps (i) and (ii) in comparative molecular phylogenetics (see text) entail the acquisition and phylogenetic analysis of molecular data. These are vast topics, well beyond the scope of this book, so only a brief introduction and some key references for interested readers are provided here.

**Molecular methods**

Many types of laboratory assay have been developed for retrieving molecular information from organismal genomes. Most of the earlier methods accessed DNA sequences indirectly, for example through assays of proteins, or via quantitative biochemical techniques such as DNA–DNA hybridization (a technique that yields numerical estimates of genetic divergence by examining the thermostabilities of nucleotide sequences). These and other tried-and-true molecular methods have been employed widely to generate phylogenetic trees as evolutionary backdrop for phylogenetic character mapping (PCM).

One of the most powerful of the modern molecular techniques – DNA sequencing – directly elucidates the precise sequences of nucleotides along specified stretches of DNA. In the past decade, refinements in laboratory methods have permitted scientists to generate large amounts of DNA sequence data, and thereby have made DNA sequencing today's most popular approach in comparative phylogenetics.

*Recommended reading*: Avise, 2002 (beginner level); Avise, 2004 (intermediate); Baker, 2000 (intermediate); Hillis *et al.*, 1996 (advanced).

**Phylogenetic methods**

Many data-analysis procedures (usually implemented in powerful computer programs) are available for reconstructing phylogenetic trees from molecular data. However, all such methods can be characterized as beginning with either: (a) numerical estimates of genetic distances among taxa (as obtained, for example, from DNA–DNA hybridization, or from tallies of nucleotide differences obtained directly by DNA sequencing); or (b) the raw character states themselves (such as specified nucleotides at many successive positions along particular stretches of DNA). In the former method, pairwise values in a matrix of genetic distances between species are "clustered" to yield an estimate of phylogeny according to user-specified algorithms (there are several available options). In the latter method, the qualitative data in DNA sequences from multiple taxa are analyzed directly to yield phylogenetic estimates based on particular assumptions or models (again there are many available options) about the nature of evolutionary interconversions among those character states.

Even with computer assistance (such as by the software PAUP (Swofford, 2000)), the search for the "best" tree is daunting when more than a few species are being compared, in part because the number of possible phylogenetic arrangements among multiple taxa is astronomical. For example, for merely 10 species the potential number of different bifurcating tree structures is more than 30 million, and for 20 taxa that number becomes about $8.2 \times 10^{21}$! From among such vast numbers of candidate trees, the objective is to identify phylogenetic arrangements that closely approximate the true evolutionary history of the taxa examined. In effect, phylogenetic algorithms in computer programs often search among possible trees for those that best comply with some user-specified evolutionary model or optimality criterion. For example, parsimony approaches (which themselves have several versions) generally operate under the assumption that the preferred tree(s) are those with the shortest total branch lengths (i.e. the fewest possible evolutionary interconversions among character states) consistent with the empirical data. (However, it should also be remembered that evolution does not always proceed along most-parsimonious routes.)

In recent years, maximum likelihood (ML) and Bayesian methods have also become popular ways to analyze molecular data and to statistically test competing phylogenetic hypotheses (see the reviews by Huelsenbeck and Rannala, 1997; Holder and Lewis, 2003). These conceptually related approaches entail computer-based explorations of tree structures (and their associated probabilities) that best explain the underlying data under specifiable models of molecular evolution. The development of fast computer software programs such as TREE-PUZZLE for ML (Strimmer and von Haeseler, 1996) and MRBAYES for Bayesian approaches (Huelsenbeck, 2000) has greatly facilitated the implementation of these newer phylogenetic methodologies.

*Recommended reading*: Hall, 2004 (beginner level); Avise, 2004 (beginner); Nei and Kumar, 2000 (intermediate); Hillis et al., 1996 (intermediate); Li, 1997 (intermediate); Felsenstein, 2004 (advanced).

of compound eyes in some insects) might each have arisen once or only a few times during evolution. Many other phenotypes (such as feather density or the number of facets in a compound eye) may vary more or less continuously, rather than qualitatively, among an array of taxa; such quantitative characteristics with numerous states often pose some of the greatest challenges for proper phylogenetic interpretation.

Throughout this book, I will use the words "characters," "traits," "features," "conditions," and "attributes" more or less interchangeably to mean multiple

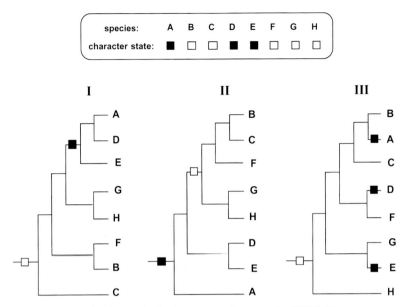

*Figure 1.3.* Introduction to the basic conceptual approach of PCM. Shown across the top are eight hypothetical species (A–H) displaying one or the other of two character states (white squares or black squares) of a particular phenotype. How these character states most likely evolved can be informed by knowledge about these species' evolutionary relationships (as can be estimated using molecular genetic data). For example, if species A–H prove to be phylogenetically related as shown in diagram I, then white-square was probably the ancestral condition for the entire group, and black-square is a shared derived condition (i.e. a synapomorphy; see Box A1) for the ADE clade. However, if the species are phylogenetically allied as shown in diagram II, then black-square was probably the original ancestral condition and white-square is a shared derived state for clade BCFGH. Many other outcomes are also possible. For example, if the true phylogeny for species A–H is as shown in III, then white-square was probably the ancestral condition from which black-square evolved independently on three separate occasions.

states of any specified phenotypes. In this generic usage, organismal characteristics to be phylogenetically mapped may encompass phenotypic descriptions of any sort (e.g. qualitative or quantitative) and at any indicated level of inclusiveness (from broad trait descriptions to those that are highly detailed). This means that biological interpretations of evolutionary outcomes will vary according to the particular phenotypes examined and questions addressed in each phylogenetic analysis.

## Phylogenetic character mapping

Step (iv) mentioned above, the primary component of comparative phylogenetics treated in this book, has sometimes been referred to as evolutionary trait analysis, comparative trait charting, or phylogenetic character mapping (henceforth PCM). Under the PCM approach, alternative states of particular phenotypic characters are matched with their associated species on an independently established phylogeny, the purpose being to reveal the evolutionary origins of those phenotypes and their probable patterns of historical interconversion. A primer to the basic methodological concept of PCM is presented in Fig. 1.3, and a somewhat fuller introduction (for the uninitiated) is provided in the Appendix. Some readers may wish to examine the Appendix first, as further technical background, before proceeding to the empirical studies presented in Chapters 2–7.

Many challenging conceptual and operational issues surround comparative phylogenetics (see Box 1.4); these too are discussed at length in an extensive scientific

---

### Box 1.4 Acknowledged limitations of comparative molecular phylogenetics

For most of the case studies presented in this book, the phylogenetic reconstructions and biological conclusions faithfully reflect those of the original authors. Thus, I have assumed that the molecular phylogenies and the PCM reconstructions were basically correct as published. This may not invariably be true, of course. Indeed, the history of comparative phylogenetics would seem to suggest that substantial fractions of published interpretations are challenged to varying degrees, sooner or later, by at least some independent researchers. The scientific sources of the resulting phylogenetic controversies can be many. Listed below are examples of hard questions that critical readers should ask before accepting the face-value conclusions from any PCM analysis.

**Does the molecular phylogeny itself correctly reflect the species phylogeny? e.g. . . .**

What types of molecular genetic assay were conducted and were they reliable?

How many unlinked genes were assayed, and how long were their sequences? [When the volume of genetic data is small, gene trees can be poor or misleading indicators of overall or composite relationships in a species tree (see, for example, Rokas *et al.*, 2003).]

Were evolutionary convergences or reversals of character states (i.e. homoplasy) likely?

Were assumptions underlying the phylogenetic analyses properly suited to the category of molecular data gathered?

Were relevant nodes in the phylogeny statistically supported and robust (e.g. did they appear consistently in multiple appropriate methods of data analysis)?

Do the molecular trees generally agree with the suspected organismal phylogenies based on conventional types of systematic evidence?

**Were the phenotype descriptions themselves adequate? e.g. . . .**

How many and what fraction of relevant extant taxa were phenotypically surveyed?

Were the phenotypes scored correctly and unambiguously?

How suitable were those surveyed phenotypes likely to have been for PCM analysis?

**Were the PCM analyses themselves properly conducted? e.g. . . .**

Were assumptions underlying the PCM reconstruction generally consistent with the known or suspected evolutionary mode of the phenotypes in question?

Were assumptions underlying the evolutionary reconstructions of sufficient flexibility that the entire PCM exercise was not flawed by circular logic?

Were inherent uncertainties in the PCM reconstructions (e.g. in the inferred character states at particular ancestral nodes) adequately captured and noted in the study's summary?

These and related questions can be difficult to answer, definitively, in a given empirical study. As a result, considerable latitude often remains for scientific controversies over particular conclusions from comparative phylogenetic analyses.

literature. The kinds of caveat listed in Box 1.4 will not be reiterated in each essay, but readers should be aware of their existence and cognizant that they can apply in varying degrees to nearly all of the case studies to be described. Nevertheless, such reservations need not unduly preoccupy readers here, for at least three reasons. First, most of the case studies were purposefully chosen, in part, because their salient findings (except where otherwise noted) were relatively straightforward and robust. Second, some of the case studies were chosen explicitly to illustrate various limitations as well as strengths of PCM approaches. Third, the biological rationales underlying current PCM efforts often remain of interest in their own right, regardless of whether particular findings are ultimately validated. In this respect, PCM-based explanations should be interpreted as reasoned but tentative hypotheses (just like those anywhere else in science): always provisional and potentially subject to reinterpretation with additional or improved evidence.

Thus, in the essays that follow in Chapters 2–7, the primary spotlights will focus on the organisms themselves and on the marvelous biological workings of nature that PCM can help to decipher.

# 2 | Anatomical structures and morphologies

Perhaps the most obvious characteristics for PCM are those readily visible to the human eye as morphological or anatomical variations among related species. Indeed, the number of published PCM studies involving macroscopic body features (such as beak shapes, limb configurations, and body morphs) probably exceeds the numbers for any other single category of organismal traits (e.g. behaviors, physiologies, or ecologies). This chapter will examine several case studies in which especially puzzling and even bizarre morphological phenotypes have been the subjects of PCM.

## Whence the toucan's bill?

About 40 species of toucan (Ramphastidae) inhabit portions of New World forests from southern Mexico to northern Argentina and Paraguay. Their most prominent feature is a colorful bill, which can be nearly as long as the bird's body. This protuberance is so outlandish that a toucan flying through a pasture or woodland can look like a silly clown pushing a banana along in front of its face. Despite its size, a toucan's bill is light and can be dexterously employed to plunder eggs or hatchlings from another bird's nest, or to pluck fruits that make up an important part of the toucan's diet. However, the spectacular colors of toucans' bills are thought to serve primarily in mate attraction or mate recognition during courtship. Thus, the toucan's bill probably evolved its impressive features via sexual selection (selection directly related to mate acquisition) as well as natural selection.

In general, avian bills often evolve rapidly, as attested by the fact that even closely related species can show widely varied bill structures. Consider shorebirds in the order Charadriiformes. Within this taxonomic group are such diverse feeding devices as: the stubby forceps-like bill of the Semipalmated Plover (*Charadrius semipalmatus*), which the bird uses to pluck small food items off mudflat surfaces; the short and slightly upturned bill of the Ruddy Turnstone (*Arenaria interpres*) for probing under shoreline pebbles; the thick wedge-shaped bill of the American Oystercatcher (*Haematopus palliatus*) for prying open the tough shells of bivalve mollusks; the long straight bills of various dowitchers, woodcocks, and snipes for probing deep into mud; the rainbow-curved bill of the Long-billed Curlew (*Numenius americanus*) for probing even deeper; and the gracefully long, upturned bills of avocets and stilts for swishing sideways through shallow waters in search of small worms and shrimp. Clearly, these varied feeding devices have arisen through natural selection.

Hummingbirds (Trochilidae) constitute another large avian group (more than 300 species) in which bill morphology varies tremendously. Bill sizes and shapes in different species often closely match the sizes and shapes of preferred flowers from which the birds sip nectar and for which they provide pollination services. For example, the Bee Hummingbird (*Mellisuga helenae*) of Cuba has a stubby little bill and sips from stubby little flowers, whereas the Sword-billed Hummingbird (*Ensifera ensifera*) of South America wields a long straight bill, lengthier than its body, to drink nectar from deep within equally elongate tubular flowers. Another South American species, the White-tipped Sicklebill (*Eutoxeres aquila*), is an example of the many hummingbirds that display strongly down-curved bills that conform nicely to the shapes and lengths of flower corollas that the birds visit.

Phenotypic traits that tend to evolve rapidly, such as birds' bills, are said to be evolutionarily "plastic," i.e. readily moldable by the particular selection pressures to which each species is exposed. Accordingly, rampant convergent evolution and frequent evolutionary reversals of state can make such traits notoriously misleading with regard to phylogenetic affinities. For example, if shape of the bill were the only criterion by which to assess phylogeny, then the Long-billed Dowitcher and Sword-billed Hummingbird might be deemed close genetic relatives, as would the Long-billed Curlew and White-tipped Sicklebill. Although any linking of particular hummingbirds with particular shorebirds would be an egregious phylogenetic mistake, the error becomes apparent only because many additional morphological (and other) characteristics of these birds indicate otherwise.

With respect to identifying toucans' closest evolutionary relatives, systematists long ago documented several morphological features (such as unique arrangements of leg tendons, and zygodactylous feet in which two toes point forward and two backward) that helped to hone the list of viable candidates down to two taxonomic groups: woodpeckers (Picidae), distributed almost worldwide and containing about 200 species; and barbets (Capitonidae), with about 90 species inhabiting the New World and Old World tropics. Both of these assemblages (as well as some others including honeyguides, jacamars, and puffbirds) were included with toucans in the order Piciformes, but it wasn't until extensive molecular data were gathered in the mid 1980s that a big surprise began to emerge.

*Figure 2.1.* Clockwise from upper left: A toucan, a New World barbet, and an Old World barbet (after Sibley and Ahlquist, 1986). Note that the structure on which these birds are perched also reflects the phylogenetic branching structure in an evolutionary tree for these three groups.

It turns out that toucans are phylogenetically allied to barbets, and in particular to barbets inhabiting the Neotropics (South America). The molecular data indicate that the toucan lineage diverged from the New World barbet lineage about 20–30 million years ago (mya), long after New World barbets had separated from Old World barbets approximately 50 mya. In other words, barbets as a whole proved to be paraphyletic with respect to toucans, meaning that toucans constitute a phylogenetic subset of the broader barbet + toucan clade (Fig. 2.1). Thus, we now know

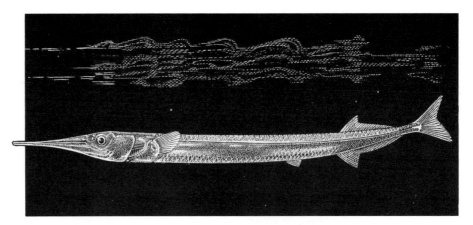

Common Needlefish

that the magnificent toucan bill had its evolutionary origin in the far more modest bill of a South American barbet-like ancestor. What remains to be better understood are the precise evolutionary forces that promoted the tremendous expansion of bill size in the toucan lineage, and, conversely, what evolutionary forces have continued to sponsor the morphological conservatism of bill size and shape in New World and Old World barbet lineages that have been physically and genetically separated for tens of millions of years.

### The beak of the fish

Adult needlefishes (Belonidae) and their close allies the sauries (Scomberesocidae) have greatly elongated snouts in which both the upper and lower jaws protrude forward from the head as needle-like extensions. The fish don't begin life that way, however. Larval needlefishes have short jaws of equal length, but as each fish matures its lower jaw elongates first, followed later by a comparable growth of its upper jaw. Thus, during the transitional phase between larva and adult, each needlefish has a long lower jaw but a short upper one. Interestingly, this intermediate state of the developing needlefish jaw closely resembles the adult condition in another beloniform family: Hemiramphidae, or "halfbeaks." In these odd-looking fishes, the lower jaw extends far forward whereas the upper jaw is merely a nubbin that looks as if it might have been chopped off in an accident or fight.

To account for how the half-beaked condition of halfbeaks and the full-beaked condition of needlefishes might have evolved from the ancestral unbeaked condition of normal fishes – including other groups of Beloniformes such as flyingfishes (Exocoetidae) and ricefishes (Adrianichthyidae) – two competing hypotheses have been advanced. Under the recapitulation hypothesis, short-jawed ancestors first gave rise to descendent halfbeaks, which in turn later gave rise to needlefishes. If so, then the observed progression of jaw structures during the development (i.e. ontogeny) of an individual needlefish would parallel the progression of evolutionary changes that transformed a short-jawed beloniform ancestor into a descendent whose adults now have long upper and lower jaws. In other words, needlefish ontogeny would mirror beloniform phylogeny, and these fishes would provide a fine example of a phenomenon first emphasized nearly 150 years ago by Ernst Haeckel in his famous "biogenetic law": ontogeny recapitulates phylogeny.

Alternatively, under the paedomorphosis hypothesis, halfbeaks were derived from ancestral needlefish stock via an evolutionary pathway in which the ontogenetic development of the jaw became arrested at the juvenile needlefish stage. Paedomorphosis, in general, is an evolutionary phenomenon in which adult descendents resemble the youthful stage of their ancestors. In effect, it is the opposite of recapitulation (in which the juvenile stage of descendents resembles the adult stage of their ancestors). As applied to beloniform fishes, the paedomorphosis hypothesis (like the recapitulation hypothesis) supposes that the original evolutionary starting condition was a short jaw, but it differs from the recapitulation scenario by proposing that subsequent evolutionary transitions were to descendent halfbeaks via ancestral needlefish-like forms.

Scientists cherish evolutionary puzzles of this sort where viable competing hypotheses make clear but opposing predictions that are empirically testable. If the recapitulation scenario is correct in the current case, then half-beaked beloniform lineages should be ancestral to full-beaked ones, whereas if the paedomorphosis scenario is correct, full-beaked lineages should be ancestral to half-beaked ones. Differentiating between these alternative hypotheses requires an explicit phylogenetic framework upon which to assess the historical polarity of morphological transformations. One such phylogeny recently emerged from analyses of nuclear and mitochondrial DNA sequences (Lovejoy, 2000).

When interpreted against this molecular backdrop (Fig. 2.2), the various jaw conditions of beloniform fishes appear most plausibly to have evolved through a general progression from short-jawed forms (as in ricefishes and other fish outgroups), to half-beaked juveniles and adults (as in most extant halfbeaks), to full-beaked adults (as in modern needlefishes). The PCM analysis further suggested that particular taxa nestled within the "halfbeak" portion of the phylogenetic tree (which proved not to be monophyletic) have secondarily lost adult long-jawed conditions on at least two separate evolutionary occasions (Fig. 2.2): in a subset of traditionally recognized "halfbeaks" that none the less have a short lower jaw, and again in the flying fishes (which also have a short lower as well as upper jaw). The main

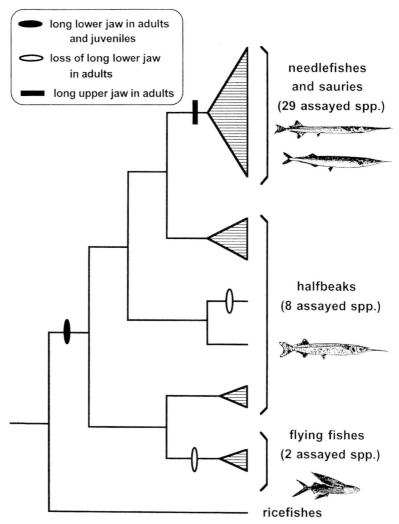

*Figure 2.2.* Simplified consensus phylogeny (based on mitochondrial and nuclear DNA sequences as well as morphological data) for beloniform fishes (after Lovejoy, 2000). Also shown are parsimony reconstructions of pertinent jaw transformations on various branches of the tree. (Note: in this and other phylogenetic drawings throughout the book, line-filled triangles are used to signify clades (i.e. monophyletic groups) within which exist multiple extant lineages (e.g. species) whose finer-scale resolution is not necessary for understanding the section's main biological points.)

conclusion, however, is that the branching topology of the molecular phylogeny favors the recapitulation scenario for the major evolutionary transitions among jaw types in beloniform fishes, and provisionally falsifies the paedomorphosis hypothesis. Further analyses (both phylogenetic and developmental) with be required to confirm mechanistic aspects of the recapitulation scenario for beloniform jaws, but it now appears quite certain that needlefishes were evolutionarily derived from halfbeaks rather than the other way around.

## Snails' shell shapes

Evolution has few inviolable truths, but scientists nonetheless try to identify evolutionary tendencies, and in the extreme have even touted some of the most consistent trends as "laws." For example, in 1893 Dollo proposed a "law of irreversibility" stating that complex adaptations, once lost, can never be regained in exactly the same form. In other words, if any complicated biological feature decays across evolutionary time (for whatever reason), it can never be precisely recouped. This law, if valid, implies a strong asymmetry or bias in the direction of evolutionary loss as opposed to the gain of particular complex adaptations. It also implies that any organisms observed to share the same complex adaptation necessarily inherited that phenotype from a shared ancestor.

Dollo's law presumes that every complex feature has an intricate if not labyrinthine genetic and developmental basis that if lost would be unlikely to re-emerge by an identical succession of causal mutations and selective events. However, biologists know of several instances in which a sudden appearance of seemingly complicated phenotypes (such as, in a fruitfly, the presence of an extra leg, or of four complete wings rather than the usual two) is due to simple mutations in homeotic genes that in effect switch the fly into a different development program (Raff, 1996). Thus, although Dollo's law presumes that many genes are modified during evolutionary changes in a complex phenotype, this may not invariably be true. Furthermore, even if a long and complex chain of genetic causation underlies an elaborate phenotype, one altered link in the chain might cause a complex phenotype to disappear, only perhaps to reappear by subsequent restoration of that one crucial link.

Dollo's law is a useful generality, but evolutionary laws seemingly were made to be broken. One interesting example involves shell architectures in gastropod mollusks (snails). Most of the approximately 35 000 species in the class Gastropoda carry a calcareous shell that is spirally coiled. The coiled shell arose early in gastropod history, and clearly accompanying this emergence was an extensive suite of phenotypic adjustments (such as the evolution of corkscrew viscera, including

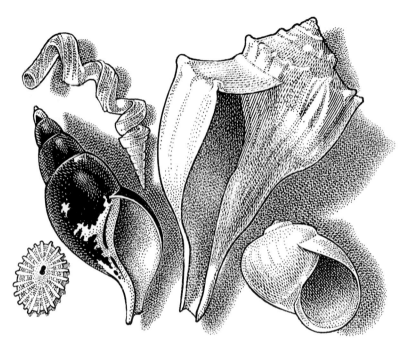

Left to right: Knobby Keyhole Limpet, Stimpson's Whelk, Fargo's Worm Shell,
Knobbed Whelk, Lobed Moon Snail

a convoluted digestive system that leads to an anal opening just above the head).
As judged by the abundance and diversity of coiled snails today, as well as across
a fossil record that extends back hundreds of millions of years, the coiled-fortress
shell design has been highly effective.

None the less, coiled shells have been lost on various occasions in gastropod
lineages. One set of examples is provided by various limpets, including hoofshells
(Hipponicidae), keyhole limpets (Fissurellidae), false limpets (Siphonaridae), and
slipper limpets (Calyptraeidae). Most of these snails have an uncoiled cap-shaped
shell that each animal pitches, like a protective tent, on a rock surface along a
pounding ocean shoreline. Another set of examples is provided by worm snails
(Vermetidae, Vamicularidae, and Siliquaridae) whose uncoiled shells grow as irreg-
ular twisted tubes. Traditionally, it was assumed that both of these types of uncoiled
shells are evolutionary dead-ends (even if highly successful in particular niches).
This supposition arose from two notions: that uncoiled shells may be ideally suited
to relatively few ecological circumstances (compared with coiled shells, whose
evolutionary lineages seem more catholic and adaptable); and that the uncoiled
condition cannot revert to a coiled state (by virtue of Dollo's law).

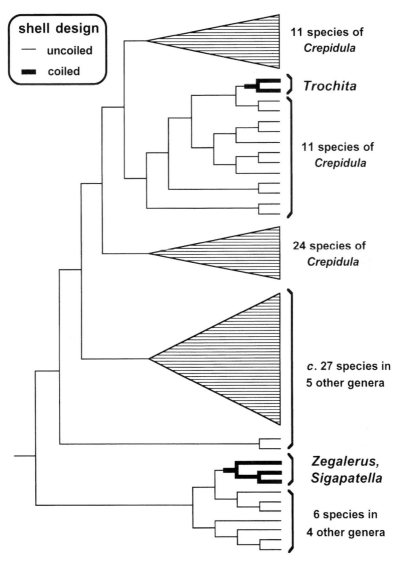

*Figure 2.3.* Molecular phylogeny for approximately 80 species in the limpet family Calyptraeidae (after Collin and Cipriani, 2003).

This latter assumption has been called into serious doubt by a recent discovery that limpets can re-evolve coiled shells after all. By examining a molecular phylogeny based on mitochondrial and nuclear gene sequences, Collin and Cipriani (2003) showed that one exceptional group of fully coiled limpets (in the genus *Trochita*) is deeply embedded within a hierarchically nested series of limpet clades that otherwise have uncoiled shells (Fig. 2.3). Members of two other exceptional limpet genera (*Zegalerus* and *Sigapatella*) likewise have coiled shells, but the more basal position of these taxa in limpet phylogeny makes it harder to eliminate the possibility that they simply retained a spiral-shell condition from an earlier limpet ancestor. If the original limpet progenitor was coiled, the most parsimonious interpretation of the phylogeny in Figure 2.3 entails a single loss and two gains of coiling in limpet history, whereas if the ancestral limpet was uncoiled the phylogeny most parsimoniously implies no losses and two evolutionary gains of coiling. Either way, instances of coiling in some species of limpet clearly reflect the re-evolution of a complex predecessor state. Coiled species are few and far between in limpet phylogeny, but they do exist, and their presence would certainly seem to violate Dollo's law.

On the outside, the coiled limpets look like their many uncoiled cousins, but cross-sections through their tented shells reveal an internal spiral architecture that at least superficially resembles the coiled and tunneled designs of most other gastropods. The genetic basis of shell coiling in limpets is unknown, so any proposed mechanism underlying the re-evolution of coiling remains conjectural at present. However, Collin and Cipriani raised an intriguing developmental possibility. They note that several limpet species, scattered across the limpet phylogeny, are known to have free-living planktonic larvae with coiled shells, but that this condition normally is not maintained into adulthood (the lifestage previously addressed in most discussions on limpet phenotypes). Thus, perhaps the coiled adult shells in *Trochita*, *Zegalerus*, and *Sigapatella* evolved by heterochrony, i.e. by a simple change in developmental timing that in this case resulted in the retention of a complex larval phenotypic state of ancestral species into the adult lifestage of descendant taxa. If so, any violation of Dollo's law in this case might be judged as merely a minor evolutionary offense, i.e. an evolutionary misdemeanor.

### More on snails' shell shapes

As described in the previous section, most gastropod mollusks have a coiled calcareous home and a twisted body to match. Like a spiral staircase that enlarges as it descends, the hollow coil of each shell wraps around a central axis or columella, beginning (developmentally and positionally) as a narrow spire and eventuating in a much broader aperture from which the snail's head and foot protrude. When viewed head-on (or foot-on), the aperture normally lies either to the right (dextral side) or to the left (sinistral side) of the columella. This establishes the snail's chirality or handedness. The vast majority of gastropods are right-handed but, interestingly, a few taxa might be described as ambidextrous, meaning in this case that dextral shells and sinistral shells both occur within the same species or taxonomic group.

Japanese land snails (from Asami *et al.*, 1998)

In several studied species, chirality is known to be controlled by alternative alleles (one dominant, the other recessive) of a single nuclear gene. The system is unusual, however, in that the genotype's expression is delayed by one generation because each offspring's phenotype is determined by its mother's (rather than by its own) genotype. This means that all full-sib or half-sib progeny from a given mother develop the same chiral phenotype, regardless of their genotypes. It also means that if a local population happened to be founded by a mutant female whose genotype specified a reversed (relative to the norm) chiral phenotype, that new population initially would consist all-at-once of snails whose handedness was opposite to that of their other predecessors.

Especially in "flat" snail species (those with low spires), handedness is critical in the face-to-face mating process. Much as two people would have difficulty shaking their partner's left hand with a right hand, two snails must normally be of the same handedness to bring their genital openings into apposition for successful copulation (see the pictures above). A dexter snail can copulate with another dexter, or a sinister with another sinister, but seldom the chiral twains shall mate. Thus, any rare dexter in a sinistral population, or any rare sinister in a dextral population, would be at a significant reproductive disadvantage because of a paucity of mating partners. So, natural selection is not even-handed with regard to handedness. Theoreticians have modeled this situation and shown that a form of frequency-dependent selection often tends to eliminate the chiral minority from any snail population that may have been polymorphic for dextral and sinistral shells.

None the less, as mentioned above, some snail lineages are collectively poly-morphic for handedness, a good example being Japanese land snails in the genus *Euhadra*. In this complex of 20 named species, four species (*quaesita*, *scaevola*, *grata*, and *decorata*) are sinistral and the other 16 are dextral. To understand the his-tories of these snails with regard to evolutionary conversions between chiral forms, Ueshima and Asami (2003) estimated a phylogeny based on mtDNA sequences. Results (according to parsimony statistics) indicated the following (Fig. 2.4): all four sinistral taxa were derived from the same sinistral ancestor; on at least three independent occasions, a named dextral species secondarily arose from a sinistral ancestor; and all lineages of the dexter *E. aomoriensis* are nested phylogenetically within the sinister *E. quaesita*.

These observations are of special interest because they suggest that speciation events in snails may sometimes come about by changes in handedness, and hence could be precipitated by genetic changes at a single locus. Because of the physical mating restrictions between dexters and sinisters, chiral reversal would consti-tute a powerful pre-mating isolating barrier underlying this mode of speciation. Perhaps a chiral-mutant mother occasionally establishes a new population, for example by colonizing the periphery of a species' range. As described above, her first-generation progeny would all have the altered phenotype, which thus would be initially common at that site. Frequency-dependent selection might then act to favor this new form of handedness and drive it to fixation, thereby quickly gen-erating a new biological species that is reproductively isolated from its progenitors.

The case involving *E. aomoriensis* is especially interesting. At least three sep-arate maternal lineages of this dextral species are nested within the matriarchal clade of its sinistral ancestor, *E. quaesita* (Fig. 2.4). One possibility (among others) is that these dextral lineages arose independently in peripheral isolates of *E. quae-sita*, under evolutionary scenarios like those described above. If so, *E. aomoriensis* would be a rare example of a polyphyletic biological species. In other words, this species would have arisen on multiple occasions from *E. quaesita*, and all three of its evolutionary lines would be reproductively compatible with one another (by virtue of sharing dextrality), yet incompatible with their *E. quaesita* progenitors (all of whom are sinistral).

### Winged walkingsticks

Dollo's law, which states that complex adaptations once lost are unlikely to be regained in evolution, has few well-documented violations, but a second example

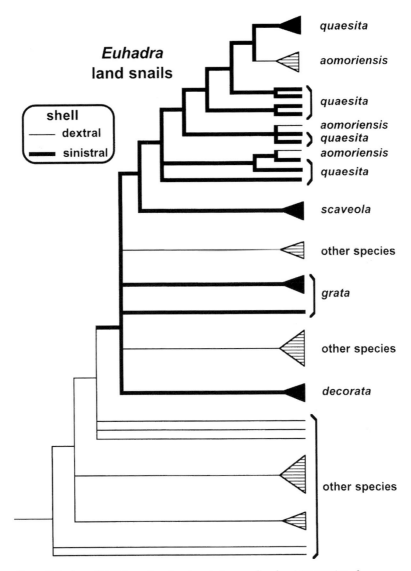

*Figure 2.4.* An mtDNA-based molecular phylogeny for about 20 species of
Japanese land snail in the genus *Euhadra*. This diagram also shows the deduced
evolutionary history of inter-conversions between left-handedness and
right-handedness (after Ueshima and Asami, 2003).

(see *Snails' shell shapes*, above, for the first) involves the loss and subsequent recovery of wings in some stick insects. Most insects are winged (pterygous), and for good
reason: active flight helps these animals escape predators and exploit untapped
resources by enabling dispersal into new habitats. However, in diverse insect orders

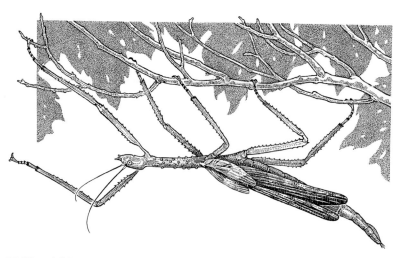

Walkingstick insect

of the class Pterygota (winged insects) at least some evolutionary lineages have become secondarily wingless. Well-known examples include fleas (Siphonaptera) and lice (Anoplura and Psocoptera). Another involves stick insects (Phasmatodea), commonly known as walkingsticks.

Stick insects are terrestrial or arboreal creatures whose cryptic bodies often escape detection by mimicking twigs or leaves. The resemblance can be nearly perfect in morphological detail, and even in behavior because many stick insects gently sway their bodies as if being a twig caressed by the wind. Among more than 3000 described stick species in three families and approximately 500 genera, about 40% are fully winged and capable of sustained flight, whereas the remainder are partially winged (brachypterous) or wingless (apterous) and are essentially grounded. Being highly cryptic affords flightless as well as flighted species a partial immunity from predation, but a special benefit of flightlessness appears to be enhanced fecundity (females in wingless species tend to lay more eggs, presumably because of relaxed restrictions on their body mass).

To understand better the historical pattern of wing evolution in Phasmatodea, Whiting *et al.* (2003) estimated a molecular phylogeny, based on mitochondrial and nuclear DNA sequences, for nearly 40 species representing most of the 19 recognized subfamilies of stick insects. Also included in their study were species representing more than 20 basal orders of winged insects that serve as outgroups to the Phasmatodea. The PCM analysis (a portion of which is summarized in Fig. 2.5) supported the notion that winglessness evolved from a winged condition near the base of the Phasmatodea clade, and that wings were later regained on perhaps four

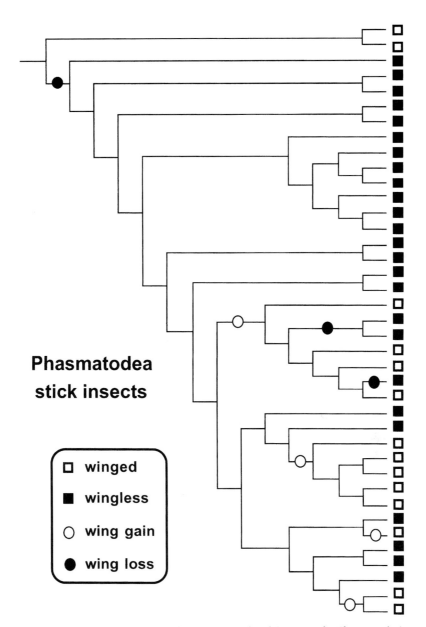

*Figure 2.5.* Molecular phylogeny for 39 species of stick insect (order Phasmatodea) onto which are mapped the probable evolutionary transitions between winged and wingless (after Whiting *et al.*, 2003).

separate occasions (and re-lost perhaps two other times). Different methods of phy-logenetic analysis yielded slightly different estimates of the numbers of secondary re-gains (and re-losses) of wings, but the unambiguous take-home message was that a complex phenotype (wings) had been evolutionarily abandoned and later recovered in particular stick insect lineages.

The etiology of this overt violation of Dollo's law differs from that described in my first example, where heterochrony appeared to be responsible for the evolutionary resurfacing of a complex phenotype (coiling in the shells of some adult snails). In the current case, involving the emergence of "flyingsticks" from walkingsticks, it seems likely that the developmental pathway for wings was genetically retained but silenced early in the stick insect lineage, only to become re-expressed in various derived lines. One plausible explanation for the long-term evolutionary retention of such an unused genetic and biochemical pathway is that the development of other walkingstick body parts might also require the same or a similar develop-mental program. In fruitflies, for example, it is known that the genetic and cellular mechanics of wing formation (involving homeotic genes and their developmental influence on primordial cell regions known as imaginal disks) are intimately related to the genetic and cellular mechanics of leg formation. Thus, it may not be too sur-prising that the basic genomic instructions for wing formation in wingless insects are conserved for long periods of evolutionary time owing to their close similarity to instructions required to form legs and perhaps other critical body structures. And, every once in an evolutionary while, as in some of the derived lineages of stick insects, the basic developmental circuitry simply gets recruited once again for wing formation.

### Hermits and kings

The coiled shells of gastropod mollusks (see above, *Snails' shell shapes* and *More on snails' shell shapes*) have profoundly affected the evolution of body designs in an entirely different group of creatures as well: hermit crabs. Whereas most arthropods in the subphylum Crustacea (lobsters, shrimp, crabs, and their allies) produce a protective casing or exoskeleton that almost completely covers the animal's body, the abdomen of a hermit crab is soft and naked. It is also asym-metrically twisted, thereby conforming nicely to the spiral chamber of a hermit's hermitage: a gastropod shell that each hermit crab has adopted as its fortified home.

Hermit crabs begin life as sea-borne nauplius larvae that look like tiny fleas, but later they settle down to take up residence in vacant snail shells. These crabs are avid house-hunters, always searching for an unoccupied dwelling with just the right size, shape, and feel. As with many human homeowners, each hermit crab continually looks to upgrade its living quarters, for example by trading up from smaller to larger dwellings as its body grows. Each time a suitable home is found, the hermit stuffs its naked little rear end into the former gastropod's shell, leaving only its legs and head protruding from its appropriated fortress.

Hermit crabs are abundant in terms of both extant species diversity and sheer numbers of individuals. On almost any tropical or temperate shoreline around the world, hordes of these animals can be seen scurrying about in their architecturally diverse travel homes, which they have salvaged from the deceased of numerous mollusk species. Hermit crabs are also common in a fossil record that extends back more than 150 million years. Thus, the adopt-a-home lifestyle has been a highly successful alternative to the construct-your-own-exoskeleton strategy employed by most other crustaceans.

The coiled shape of the hermit body is unusual among crabs, most other forms of which possess a straight abdomen and a tail that typically remains safely tucked under the animal's thorax. Oddly, however, king crabs in the family Lithodidae have an asymmetrically twisted abdomen reminiscent in shape to that of hermits. This had long led to suspicions that king crabs and hermit crabs might be closely related, despite their obvious differences in size (king crabs are huge) and lifestyle (king crabs inhabit cold deep seas and they don't house themselves in snail shells). However, it was not until DNA sequence data became available that the phylogenetic tightness of king and hermit crabs was confirmed. Indeed, using mtDNA gene sequences, Cunningham *et al.* (1992) showed that king crabs occupy one branch deep within the broader phylogenetic tree of hermit crabs (Fig. 2.6). This discovery appears to solve the mystery of why free-living king crabs have a twisted abdomen. Evidently, this body condition is simply a phylogenetic legacy retained from snail-shell-inhabiting hermit crab ancestors. Furthermore, molecular-clock considerations led Cunningham and colleagues to conclude that the evolutionary transition from hermit crabs to king crabs probably took place about 13 to 25 million years ago.

These PCM findings beg the question: what originally prompted creatures in the king crab lineage to abandon the safety of snail shells? No one knows for sure, but one reasonable hypothesis is that snail shells are relatively scarce in the deep-sea habitats where king crabs live. Without the assurance of safe havens, selection pressures must have been strong for the hermit-like ancestors of king crabs to revert to more normal crab-like phenotypes by covering their otherwise vulnerable abdomens in a protective exoskeleton. So, the evolutionary rationale for body fortification in king crabs is much like the conventional scenario that applies to typical crabs, but with a little added twist.

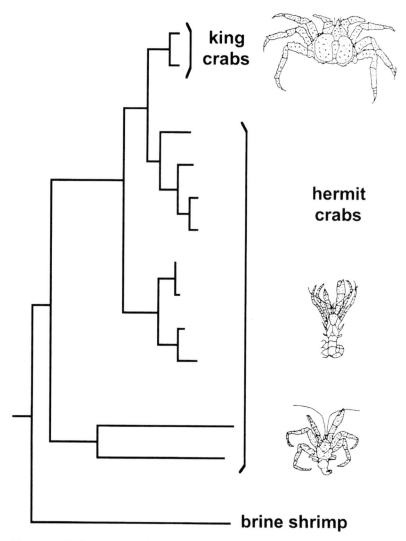

*Figure 2.6.* Phylogenetic tree for hermit crabs and king crabs (plus brine shrimp as an outgroup) based on ribosomal gene sequences in the mitochondrion (after Cunningham *et al.*, 1992).

### True and false gharials

Although molecules and morphology sometimes paint very different phylogenetic pictures for a given taxonomic group, overt conflicts between these two sources of information are in truth relatively uncommon. Disharmonic outcomes tend to be emphasized in the scientific literature (including this book) because they are curiosities that attract special interest. This section describes one such seemingly

True Gharial

blatant inconsistency between the phylogenetic trees generated from molecular and morphological data. This case involves two species of gharial (also known as gavials), in the order Crocodilia, which also includes 21 living species of crocodile, alligator, and caiman.

Gharials look much like other crocodilians except that they have narrow elongated snouts. The True Gharial (*Gavialis gangeticus*) inhabits rivers of the northern Indian subcontinent; the False Gharial (*Tomistoma schlegelii*) lives in freshwater swamps, lakes, and rivers of Indonesia and Malaysia. As reflected in its common name, the False Gharial traditionally was classified not as a genuine gharial (Gavialidae), but rather as a crocodile (Crocodilidae) whose ancestors supposedly had converged in general appearance on the true gharial by independently evolving a long slender muzzle. This conclusion, based on detailed appraisals of numerous other morphological traits (and cladistic reasoning) had convinced most herpetologists that despite superficial outward appearances, the False Gharial and True Gharial were not close evolutionary relatives. This conventional wisdom is summarized in Figure 2.7A.

An entirely different view emerged later from phylogenetic analyses of mtDNA sequences. According to this new molecular evidence, the False Gharial had incorrectly been assigned to the Crocodilidae, and instead constituted a close sister lineage of the True Gharial (as shown in Fig. 2.7B). Thus, there seemed to be a flagrant discord between molecules and morphology with regard to the phylogenetic placement of True and False Gharials in the crocodilian tree.

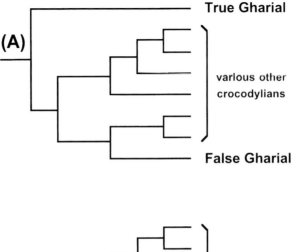

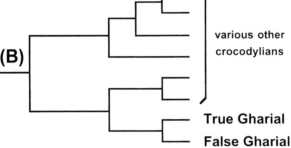

*Figure 2.7.* Two divergent and competing views regarding the phylogenetic placements of True and False Gharials within the crocodilian clade (after Harshman *et al.*, 2003). (A) The traditional scenario based on phylogenetic analyses of many morphological characters. (B) An alternative scenario based on phylogenetic analyses of numerous molecular characters.

Because all loci in the mitochondrial genome are genetically linked and evolve as a unit, an mtDNA phylogeny is a quintessential example of a gene tree. Theory has shown that the topology of any gene tree (mitochondrial or nuclear) can differ from the consensus topology of a species tree for any of several reasons ultimately relating to genomic sampling errors. In effect, many gene trees comprise a species tree, so any species' cladogram really should be thought of as a statistical "cloudogram" (Maddison, 1997) of quasi-independent gene phylogenies. This means that inferring a species tree from any single gene tree can be problematic. So, one possibility is that the mtDNA-based phylogeny for crocodilians is in gross error and the morphology-based estimate is correct. Alternatively, the morphology-based estimate might be in error and the mitochondrial phylogeny correctly reflects the species' tree. A third possibility is that mitochondrial and morphological traits both correctly record crocodilian phylogeny but that one or the other data set was improperly interpreted. A fourth possibility is that both available data sets imply an incorrect topology for the crocodilian tree.

Deciding between such possibilities demands additional genetic information and further analyses. In this case, various classes of molecular data were also gathered from the nuclear genome; they all tended to support the mtDNA-based phylogeny by indicating a close sister-group relationship between the False Gharial and True Gharial (as shown in Fig. 2.7B). One key study (by Harshman *et al.*, 2003) involved a detailed molecular-phylogenetic analysis of the *c-myc* proto-oncogene housed in cell nuclei. The authors deemed the molecular evidence now to be conclusive. They also raised a possibility that the phylogenetic discrepancy between the molecular and morphological data might be more apparent than real if (as they suspect) some of the previously employed morphological characters had been coded or interpreted incorrectly. For example, the long narrow snout that is the unique hallmark of the gharials would itself now appear (in light of the molecular phylogeny) to be a shared derived trait rather than a product of convergent evolution.

If all of this is indeed correct, it means that earlier morphology-based appraisals had falsely denied rightful sister-species status for the False Gharial and True Gharial. Thus, the False Gharial is probably a genuine gharial after all, although it cannot actually be called a True Gharial because that name is already taken.

## Loss of limbs on the reptile tree

As their name implies, tetrapods (vertebrates other than fishes) typically have four limbs, but this is not invariably true. Within the order Squamata ("scaly-skinned" lizards, snakes, and allies), legs are sometimes absent or severely reduced. For example, worm-lizards (suborder Amphisbaenia) are burrowing reptiles that use their legless rope-like bodies and wedge-shaped heads to force themselves through the soil. Glass lizards (often placed in the family Anguidae) are another limbless lot that superficially resemble large (albeit scaly) worms. True snakes (suborder Serpentes), with approximately 2700 living species, are the most ubiquitous and conspicuous of the limbless reptiles. Among the features that distinguish extant snakes from other legless squamates are deeply forked tongues and an absence of eyelids and external ears.

To estimate squamate relationships, Vidal and Hedges (2004) gathered DNA sequences from two slowly evolving nuclear genes. Results, summarized in the left panel of Figure 2.8, indicate that living snakes are monophyletic, as are several other traditionally recognized squamate groups such as iguanid-like lizards. This gives added confidence to the general branching structure of this molecular tree.

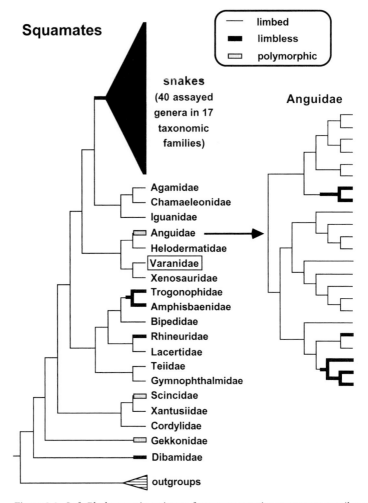

*Figure 2.8.  Left*: Phylogenetic estimate for representative squamate reptiles and outliers based on nuclear gene sequences (after Vidal and Hedges, 2004). *Right*: Molecular phylogeny estimated from mtDNA sequences for 23 species in the lizard family Anguidae (after Wiens and Slingluff, 2001). Both the coarse-level (left) and finer-level (right) phylogenies suggest multiple evolutionary origins of limblessness in squamates.

From morphological and other evidence, the ancestral condition for Squamata was "limbs present"; the same can be said for specific clades such as Serpentes and Anguidae within Squamata (see right panel in Fig. 2.8). Thus, the molecular phylogenies indicate that leglessness is a derived state that originated several times independently in squamates. These and other evolutionary instances of limb loss or severe limb reduction in Squamata have typically been associated with increases

in body length relative to body diameter, and with pronounced shifts toward undu-
latory locomotion.

The study by Vidal and Hedges was motivated by a scientific controversy
surrounding two competing hypotheses for the evolutionary origins of snakes.
A terrestrial scenario posited that snakes are descended from ancestral squamates with burrowing or semi-burrowing habits. The exact lineage is not necessarily specified, but the general idea is that the reduction and eventual loss
of limbs occurred in some ancestral squamate that had adopted a fossorial
lifestyle. Under a marine scenario, by contrast, snakes originated from water-
dwelling ancestors. Both the terrestrial and marine hypotheses have been interpreted as consistent with various morphological details shared by living snakes
and their hypothesized land-dwelling or water-inhabiting ancestors, but of course
both scenarios cannot simultaneously be correct (assuming that snakes are
monophyletic).

A marine origin for snakes is not as implausible as it might seem. Today, about
50 species of venomous snake (family Hydrophidae) inhabit Indo-Pacific waters.
Furthermore, a major scientific splash was made with the recent discovery of fossil remains from extinct sea snakes (pachyophiids) that had small but evident
hindlimbs. However, water-inhabiting snakes (fossil or modern) certainly do not
clinch the case for a marine scenario of snake origins, because these animals
could also have arisen from terrestrial snake ancestors that secondarily invaded the
sea.

As typically presented, the marine hypothesis actually posits that fossil and
living snakes evolved from Cretaceous marine reptiles known as mosasaurs. No
mosasaurs are alive today, but they are survived by presumptive relatives in the
family Varanidae (the impressively large monitor lizards). Thus, if the conventional
marine hypothesis is correct, mosasaurs and extant snakes should be most closely
related, followed by varanid lizards as the immediate sister group to this pair. In
other words, snakes should be highly derived varanid-like lizards. In the molecular
phylogeny from Vidal and Hedges, however, varanids proved to be neither ancestral to nor the immediate sister group of Serpentes (Fig. 2.8). Thus, according to
the authors' interpretation, the hypothesized evolutionary chain of transition from
mosasaurs to early marine snakes appeared to be broken; accordingly, the original
marine hypothesis for snake origins was provisionally refuted.

Given the back-and-forth nature of the debate over the land-first versus sea-first
habits of ancestral serpents, I doubt that the molecular findings described above
will be the final word on the matter. And, even if the terrestrial scenario for snake
ancestry is correct, much remains to be learned about snake's precise evolutionary
progenitors and their morphologies and lifestyles. In any event, what is most clear

from squamate phylogeny is the apparent readiness with which various reptilian groups have relinquished the locomotory limbs that most tetrapods find essential.

## Fishy origins of tetrapods

About 400 million years ago, amphibian-like fishes slowly crawled out of primordial seas or lakes to become the world's first tetrapods (land vertebrates). The fossil record indicates that the fleshy pectoral and pelvic fins of these "lobe-finned" fishes (subclass Sarcopterygii) were quite unlike those of their ray-finned relatives (Actinopterygii). These lobed fins had complex internal bony structure, and thus looked and operated like incipient limbs. The amphibians into which these remarkable fishes gradually transformed soon began to diversify, with some lineages eventually leading to reptiles (particular subsets of which later spawned all mammals and birds). Thus, in a phylogenetic sense, all tetrapods could justifiably be considered modified sarcopterygian fishes.

The piscine sarcopterygians themselves are (or were) a diverse lot, as best exemplified by ancient fossil groups sometimes referred to collectively as rhipidistians. Most lineages and species of sarcopterygian fish went extinct hundreds of millions of years ago, but a precious few are still alive today. First are the lungfishes (Dipnoi), with a total of about six extant species on three southern continents (Australia, South America, and Africa). Of these, the Australian lungfish (*Neoceratodus forsteri*) is of special interest because it most closely resembles (at least superficially) the general body plan of the supposed tetrapod ancestor. The other extant sarcopterygian fish is the famous coelacanth, initially named *Latimeria chalumnae* (but which may in fact be a complex of two or more closely related species in the Indian Ocean and adjacent waters). Coelacanths were formerly thought to have gone extinct about 65 million years ago, so their rediscovery off the east coast of South Africa in 1938 thrilled the scientific world.

The extant coelacanths and lungfishes are biological treasures because they provide living examples of how the ancient tetrapod ancestor might generally have looked and acted. They also carry intact DNA suitable for phylogenetic analyses. This new goldmine of molecular information has generated great interest in the question of whether the closest living relatives of tetrapods are lungfishes or coelacanths. Actually, three distinct hypotheses exist (Fig. 2.9): lungfishes are the sister group to tetrapods; coelacanths are the sister group to tetrapods; and lungfishes plus coelacanths are the sister clade to tetrapods.

Morphological evidence is equivocal regarding these competing possibilities. Several arcane anatomical details (such as free hyomandibular bone, presence of a glottis, internal nostrils, truncus arteriosus of the heart, and joined pelvic girdles)

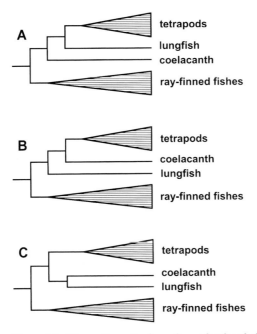

*Figure 2.9.* Three alternative hypotheses for the phylogenetic root of tetrapods (based on diagrams in Meyer and Wilson, 1990).

would seem to unite lungfishes and tetrapods to the exclusion of coelacanths. Other anatomical features (such as an endolymphatic commissure) characterize coelacanths and tetrapods but not lungfishes. Still others (such as electroreceptors mostly on the snout, and a septum dividing the hemispheres of the brain) unite lungfishes and coelacanths to the exclusion of tetrapods. All of the morphological features listed above are also absent (or displayed in an alternate state) in the ray-finned fishes, so at face value each would seem to be a shared derived rather than an ancestral trait for the taxa that possess them. However, not all of the traits listed above can be valid synapomorphies because the clades they provisionally characterize are in disagreement. Ergo the phylogenetic dilemma that molecular data might hope to resolve.

A wealth of genetic information (complete nucleotide sequences of mtDNA genomes as well as DNA sequences from numerous nuclear loci) have since been gathered and analyzed for coelacanths, lungfishes, and various ray-finned fishes and tetrapods. Surprisingly, the current results are less than definitive, although they do perhaps give mild support to hypothesis "A" in Fig. 2.9. Thus, among living fishes, the lungfishes may be phylogenetically slightly closer to land vertebrates that are the other candidates. Regardless of the true branching structure of the phylogeny, the molecular data strongly suggest that the three internal nodes in evolutionary tree in Fig. 2.9 must have been quite closely spaced in evolutionary time.

In addition to being of academic interest, these phylogenetic analyses of tetrapod origins illustrate a broader point about PCM. Namely, the proper interpretation of morphological character-state evolution can depend critically on the exact structure of a molecular phylogeny. In this case, if lungfishes truly are the sister group to tetrapods, then the presence of a glottis, joined pelvic girdles, and various other traits (some of which were listed above) are probable synapomorphies that originated after the separation of the lungfish–tetrapod ancestor from earlier fish stock. Taken at face value, this also would mean that several other phenotypic characters are phylogenetically misleading. Thus, for example, "electroreceptors mostly on the snout" and "septum dividing hemispheres of the brain" would not define a lungfish–coelacanth clade. Perhaps such shared traits arose in a precursor of sarcopterygians but were secondarily lost in a more immediate ancestor of the tetrapod clade, or perhaps these features were gained independently in the lungfish and coelacanth lineages.

Another possibility is that a different phylogenetic arrangement in Fig. 2.9 is correct, in which case all of the provisional PCM interpretations listed above would again have to be revisited. The broader point, illustrated by sarcopterygian fishes, is that wholesale reinterpretations of how phenotypic traits evolved are sometimes necessitated by even slight alterations in the underlying topology of a phylogenetic tree.

### Panda ponderings

Some creatures display such peculiar assortments of anatomical and behavioral features that the phylogenetic challenge is simply to identify the major taxonomic groups to which they are most closely related. For nearly 140 years, scientists have pondered phylogenetic puzzles provided by the world's two species of panda: the Giant Panda (*Ailuropoda melanoleuca*) and the Lesser or Red Panda (*Ailurus fulgens*), both native to China. Although both species are herbivorous, little doubt exists that they belong to the taxonomic order Carnivora (carnivores). That, however, is where the certainty ended.

In 1869, a French missionary and naturalist in China (Père Armand David) provided the first scientific description of the Giant Panda, naming it *Ursus melanoleuca*, which means "black-and-white bear." The creature does look superficially like a bear (family Ursidae), but it also displays many traits that are not at all bear-like: flattened teeth only (i.e. no canines), a strictly herbivorous diet (of bamboo), lack of hibernation, a plaintive bleating voice (like a sheep's), and an opposable "thumb" (actually a modified wrist bone) that permits the animal to grasp small branches. David sent skeletal material from the Giant Panda to a scientific colleague (Alphonse Milne-Edwards), who concluded that the animal was instead more closely related to the Red Panda, which at that time was considered to belong to the raccoon family (Procyonidae). Accordingly, Milne-Edwards changed the generic assignment of the Giant Panda from *Ursus* to *Ailuropoda*. But this was far from the end of the story. More than a century later, and despite the appearance of more than 40 morphological treatises on the subject, no clear scientific consensus had yet emerged regarding the Giant Panda's ancestry.

Phylogenetic relationships of the Red Panda have been no less controversial. Debates have swirled as to whether this species is related most closely to the Ursidae, Procyonidae, Mustelidae (weasels, otters, badgers, wolverines), Musteloidea (procyonids + mustelids, including skunks), or Pinnipedia (seals, sea lions, walruses). All of these taxa are traditionally placed within (or, in the case of Pinnipedia, phylogenetically adjacent to) the superfamily Arctoidea. Indeed, the only groups of extant Carnivora that have not been seriously entertained as the Red Panda's closest kin are members of the somewhat more distantly related superfamilies Canoidea (dogs, wolves, foxes, and allies) and Feloidea (cats, civets, mongooses, and hyenas).

It wasn't until the mid 1980s that the pandas' historical genetic relationships finally began to clarify. The cleanest phylogenetic signal has come from the Giant Panda. Based on a wealth of molecular data from an exceptional variety of laboratory techniques (including protein analyses, immunological comparisons, DNA hybridization, and direct DNA sequencing of several genes), a consensus emerged: the Giant Panda represents a hereditary line that separated from proto-bears early in the history of the ursid lineage, perhaps some 20 million years ago. In other words, the Giant Panda is indeed a "bear," albeit one whose ancestors branched off very early in ursid evolution. An emerging picture for the Red Panda is not quite so clear despite a growing body of molecular data. None the less, at least two formerly viable possibilities seem to have been firmly rejected: that the Red Panda is a sister species to the Giant Panda; and that the Red Panda is part of the ursid clade. Instead, this species almost certainly belongs within the Musteloidea.

Molecular findings for the two panda species have been put together to provide a provisional composite phylogeny (Fig. 2.10) for carnivores that traditionally had been placed in the Arctoidea (for simplicity, the position of the Pinnipedia is not included in this diagram). The musteloid clade is clearly distinct from the ursid clade, and itself contains at least three major sublineages: skunks, procyonids + mustelids, and the Red Panda. Statistical analyses indicate that the exact branching

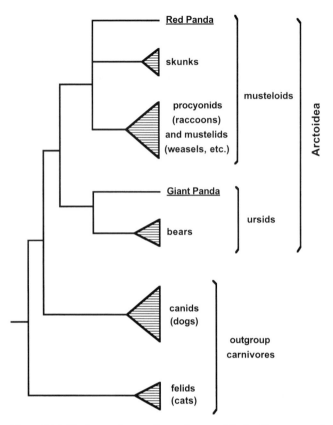

*Figure 2.10.* Phylogeny for pandas and some of their allies as recently estimated from molecular data (after Flynn *et al.*, 2000).

order of these three sublineages is not yet definitively resolved with molecular data, so for now this portion of the phylogeny is shown as having an unresolved trichotomy.

When the anatomies and behaviors of the Giant Panda and Red Panda are considered in the light of this new phylogenetic arrangement for Arctoidea, it becomes apparent that the use of the same colloquial name (panda) is misleading. It also seems clearer that various morphological features and behaviors unique to the Giant Panda, and others unique to the Red Panda, are independent autapomorphies that evolved, respectively, along two lengthy and distinct branches of the Arctoidea tree. Thus, with the benefit of molecular phylogenetic hindsight, it becomes less mysterious why these traits formerly generated such phylogenetic uncertainties. Any highly idiosyncratic trait that is confined to a single evolutionary line may offer few, if any, clues about phylogenetic links to other lineages that lack the trait.

### Fossil DNA and extinct eagles

Molecular appraisals of phylogeny normally rely on DNA sequences extracted from living species. In special circumstances, however, it is also possible to recover and sequence DNA from well-preserved fossils. Normally when an animal or plant dies, its DNA degrades rapidly, but under exceptional conditions of rapid desiccation, low-temperature freezing, or high salt concentration, some DNA molecules may remain relatively intact for as long as about 100 000 years (in the extreme). With the advent of a powerful laboratory technique known as the polymerase chain reaction (PCR), scientists can now extract, amplify, and analyze short DNA sequences from fossil material such as desiccated mummies, frozen carcasses, salt-preserved museum skins, or the marrow of intact bones. Indeed, studies of ancient DNA or "fossil DNA" have spawned a whole new discipline: molecular paleontology.

In some cases, phylogenetic analyses of fossil DNA have yielded evolutionary insights that could not have emerged from molecular examinations of living species alone. Michael Bunce and his colleagues (2005) provided a fine example in their recent genetic study of one of the largest flying birds the world has ever known: New Zealand's extinct Haast's Eagle (*Harpagornis moorei*).

New Zealand has been a particularly fascinating evolutionary theatre owing to its diverse topography and longstanding physical isolation from Australia and other continents of the Southern Hemisphere. For example, the complete absence of native predatory mammals in New Zealand was undoubtedly a key factor in promoting the evolution on the islands of a wide variety of flightless birds (most now extinct), ranging from a tiny wren to non-flying ducks, rails, wattled crows, and kiwis, to gigantic moas (some of which were three times the weight of an ostrich). Among the wingless moas alone, approximately 20 species inhabited the islands before the last of them became extinct following the settlement of New Zealand by Polynesian peoples about 700 years ago. Moas were the islands' top-echelon herbivores, generally occupying the ecological niches of large herbivorous mammals (such as kangaroos or deer) that inhabit continental landmasses.

Haast's Eagle was another of New Zealand's gigantic birds, with an adult body mass of 10–15 kg, a wingspan of 2–3 m, and talons each more than 10 cm long. It apparently specialized on moa prey, striking and then gripping a moa's pelvic area with one foot while striking its head or neck with the other. It is estimated that the massive claws of *Harpagornis moorei* could penetrate about 5 cm of skin and flesh to pierce and crush underlying bones as thick as 10 mm. Haast's Eagle, like its moa prey, became extinct following arrival of the Polynesians in the thirteenth century.

Previously, some ornithologists supposed that Haast's Eagle was related most closely to another hefty species – the Wedge-tailed Eagle (*Aquila audax*) – that still lives in Australia today. Although only about one-third the mass of *Harpagornis moorei*, the Wedge-tailed Eagle is among the larger of extant eagle species, and from this and other evidence (such as geographic distribution) it seemed a likely candidate for representing a sister lineage to that of New Zealand's Haast's Eagle. To test this hypothesis, Bunce and colleagues (2005) extracted mtDNA molecules from the fossil bones of two specimens of *Harpagornis moorei*, and compared their nucleotide sequences to those that had been gathered from extant species of eagle from around the world. Phylogenetic analyses of these data yielded a surprise: the closest living relatives of *Harpagornis moorei* appeared to be the Little Eagle (*Hieraaetus morphnoides*) and Booted Eagle (*Hieraaetus pennatus*) of eastern Asia, whereas *A. audax* was only a distant evolutionary cousin (Fig. 2.11).

Members of the genus *Hieraaetus* are actually among the smallest living eagles, having less than one-tenth the mass of *Harpagornis moorei* and only half the wingspan. Although at first glance they might seem unlikely candidates for being the Haast's Eagle's closest kin, the molecular genetic data indicate otherwise. Thus, rather than having evolved directly from a large-bodied ancestor, it now seems probable that *Harpagornis moorei* evolved from much smaller eagles that may have colonized New Zealand about 1–2 million years ago (based on the observed magnitude of mtDNA sequence divergence between *Harpagornis* and *Hieraaetus*).

If this evolutionary scenario is correct, then the rate and magnitude of body enlargement in the *Harpagornis* lineage is unrivalled in the avian world. Presumably, the absence of mammalian predators on New Zealand, coupled with the large body sizes of potential prey (and perhaps intense competition from other hawks for smaller food items), provided the selective pressures that resulted in the rapid evolutionary enlargement of body size by Haast's Eagle, a truly awesome aerial killer.

### The Yeti's abominable phylogeny

The Himalayan Yeti, also known as the Abominable Snowman, has a storied history. Perhaps the first "reliable" report of a Yeti came in 1925 when N. Tombazi, a Greek photographer, sighted an ape-like creature moving across a Himalayan slope at about 15 000 foot elevation. Unfortunately, Mr Tombazi failed to photograph the animal. Across the ensuing decades, other Himalayan trekkers have likewise reported close encounters with Yetis, or at least with their monstrous footprints in the snow. Such evidence has given rise to speculation that the Yeti is a close evolutionary cousin of North America's equally mysterious Bigfoot (or Sasquatch).

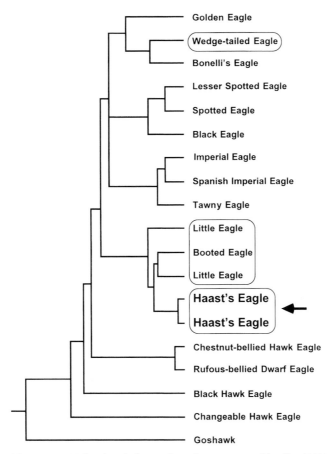

*Figure 2.11.* Molecular phylogeny based on extant and fossil mtDNA sequences for 16 eagle species plus a Goshawk outgroup (after Bunce *et al.*, 2005). Note the placement of the extinct giant Haast's Eagle in a clade otherwise composed of relatively small-bodied taxa.

Other members of this reclusive family may be the Mapinguari of the Amazon, and the Yowie of Australia.

Actually, Sherpa legend has it that the Himalayan Yeti may be any of three distinct animals: large shaggy creatures (perhaps the Tibetan blue bear), "thelma" (a type of gibbon), and "min the," the "true" Abominable Snowman. The latter is usually presumed to be some form of great ape, evolutionarily related to chimpanzees, gorillas, or early humans. Generally consistent with this notion are descriptions of the true Yeti as "a hairy, reddish-brown creature with a rigid crown" (Matthiessen, 1979), or "A sort of enormous monkey . . . with a huge head like a coconut" (Hergé, 1960). Unfortunately, determining the exact phylogenetic position of the Yeti has been

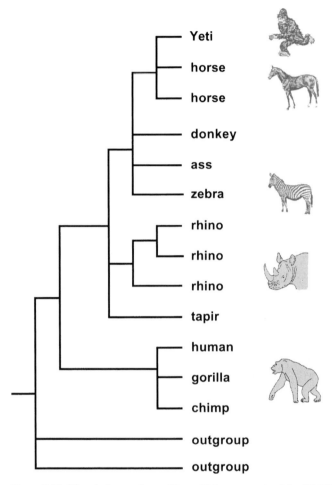

*Figure 2.12.* The phylogenetic position within mammals of the "Yeti" Himalayan hair sample, as inferred from comparative sequence analyses of mitochondrial DNA (after Milinkovitch *et al.*, 2004). Each terminal node in this tree represents the DNA sequence from one assayed specimen.

difficult because, despite vast field effort and numerous expeditions, no bones, no scat, no artifacts, and no dead bodies have ever been discovered. Having had nothing more tangible than fleeting footprints to work with, some scientists have questioned the true Yeti's existence, much to the consternation of true believers.

It thus came as a welcome event when Yeti hairs were discovered in 1992, and made available for DNA sequence analysis. Peter Matthiessen and Thomas Laird had stumbled upon these hairs while exploring a remote Himalayan region near the Tibetan border, in the mysterious Kohla Valley outside the city of Lo Monthang.

They had noticed Yeti footprints in the snow, and nearby they recovered twisted hairs that were identified by their local guides as clearly being of Yeti origin. These precious hair samples were then sent to a molecular forensics laboratory in Belgium for phylogenetic analysis of their DNA contents.

There, a team headed by Michel Milinkovitch successfully recovered and assayed ribosomal (r) DNA sequences from the "Yeti" and compared results to previously gathered rDNA sequences from numerous other mammal species. To their dismay, the rDNA sequences from the Yeti hairs closely resembled those of ungulate animals (horses in particular), rather than primates (Fig. 2.12). Thus, genetically and evolutionarily, the Yeti was basically a horse. From this unambiguous phylogenetic finding, the researchers drew (intentionally) what might best be described as an abominable conclusion: The true Yeti, being in the horse clade, presumably must have evolved extensive morphological similarities to apes, making this an astonishing example of evolutionary convergence at the phenotypic level (see Fig. 2.12). In other words, with tongues firmly in cheeks, the authors purposefully chose to overlook the more obvious or parsimonious explanation (that the unknown hair simply had been shed by a horse).

Milinkovitch and coauthors published their paper in a high-quality scientific journal (*Molecular Phylogenetics and Evolution*) on April 1, 2004 (April Fool's Day). The Himalayan hair sample truly had been collected by people convinced of its authenticity, and the molecular forensic analysis was indeed *bona fide*. However, in a spirit consistent with the longstanding tradition of non-skeptical inquiry into Yeti origins, the authors erected their convergent-morphology scenario as a spoof.

To be evenhanded, the authors also poked a bit of fun at overly serious scientists who not infrequently become embroiled in heated disputes over the phylogenetic positions of particular taxa. In a refreshing footnote to their article, Milinkovitch and his colleagues (2004) reminded evolutionary biologists that they too sometimes "need to retain a sense of humor in their efforts to reconstruct phylogenetic relationships."

# 3 | Body colorations

For creatures that emit or receive visual cues, body colors (or lack thereof) often serve key communication roles. In the context of predation, for example, conspicuousness in the form of blatant warning colorations can be a key to individual survival for both toxic prey organisms and their potential predators, whereas crypsis usually benefits prey that are highly palatable. In the context of sexual communication also, body colorations can influence genetic fitness through their rather direct impacts on reproductive success. Brightly colored males, for example, may tend to acquire more mates than their drab competitors, and hence be evolutionarily favored by sexual selection (see several sections below).

Colors are sensations of light induced in the nervous system of beholders by electromagnetic waves of various frequencies. In terms of biological effects, colors can be thought of as functional outcomes of mechanistic interactions between a transmitter and a recipient, so different observers may perceive the same object differently (for example, many pollinating birds and insects are highly attuned to ultraviolet flower colors that are invisible to people and most other mammals). However, colors also can be interpreted as the electromagnetic wavelengths themselves, in which case they become properties of the light source and the transmitting organism's reflective surfaces (irrespective of observer perceptions). In that sense, too, reflected wavelengths *per se* become another aspect of an organism's external phenotype. Regardless of how they are "viewed," body colors have been the subject of a sufficient number of PCM analyses to warrant a separate chapter here.

## Light and dark mice

Often, particular evolutionary hypotheses are put to critical test by comparing results of PCM analyses conducted jointly on different suites of characters. A case-in-point involves Rock Pocket Mice (*Chaetodipus intermedius*), which inhabit rocky outcrops in deserts of the American Southwest. This case study will also introduce the notion that PCM can be conducted even at the micro-evolutionary scale of conspecific (same-species) populations.

These pocket mice come in two basic pelage tints: light and dark. The light, sandy-colored mice live mostly on light-colored substrates, whereas the dark (melanic) mice are found on blacker rocks that were formed from ancient flows of dark-hued basaltic lava. Thus, the two pelage types map almost perfectly onto like-color habitats that the animals occupy. These matches between fur and background help to conceal the mice and afford them greater protection from visually oriented avian and mammalian predators. Owls, for example, are important predators on mice, and they have been experimentally shown to discriminate finely between light and dark prey even at night. Thus, the cryptic pelages of pocket mice clearly have been shaped by natural selection.

Dark lava flows are surrounded by lighter desert habitat and usually are isolated from one another by hundreds of kilometers. The possibility thus exists that melanic mice on separate lava flows arose and evolved independently, perhaps via different genetic mechanisms, from widespread light-furred ancestors. This hypothesis was put to a critical test in molecular analyses comparing specific coat-color genes (probably under strong selection) with those from functionally unrelated genetic markers (some of which might be neutral, i.e. invisible to natural selection).

In laboratory mice (mostly *Mus musculus*), approximately 80 nuclear genes have been identified that affect major categories and finer nuances of pelage color variation in rodents. Two of these genes, in particular, seemed likely candidates for functionally influencing the specific category of light versus dark hair shades in the Rock Pocket Mice. One gene encodes melanocortin-1-receptor (MC1R), a protein highly expressed in melanocytes (pigment cells containing melanin granules), and the other encodes an agouti-signaling protein ("agouti") that acts antagonistically to MC1R by reducing the production of brown and black melanins.

Hopi Hoekstra, Michael Nachman, and colleagues first isolated and DNA-sequenced both of these genes in pocket mice inhabiting several locales with light and dark substrates in New Mexico and at one site (Pinacate) in Arizona (Nachman *et al.*, 2003). Molecular variation in the agouti gene proved to bear no relation to light versus dark pelages. However, at the Arizona site, but not in New Mexico, specific mutations at the MC1R gene proved to be in perfect association with light versus dark coats. These findings strongly suggested that MC1R is the direct causal basis of pigmentation differences in the Arizona mice but not in the New Mexico animals, and hence that adaptive melanism probably arose at least twice independently in the evolutionary history of *C. intermedius*.

One caveat to this conclusion is that the correlation at the Arizona site between coat colors and specific MC1R alleles (alternative forms of a gene) might be spurious rather than indicative of a causal relationship between the two. At least in theory, specious correlations of this sort might arise from historical accidents of association between functionally unrelated traits, as could occur if a population has been strongly subdivided genetically. (For example, alleles for blond hair and blue eyes tend to be correlated in human populations not because they are causally connected, but rather because they both stem historically from peoples that inhabited Scandinavian regions.) Counter-wise, in any random-mating population (or set of populations with high rates of genetic exchange), strong correlations between particular phenotypes and genes are theoretically expected only if those genes (or others linked tightly to them on the same chromosome) are truly causally responsible for the phenotypic variety.

To test for pronounced population subdivision at the Pinacate site, the team led by Hoekstra and Nachman also surveyed pocket mice for mitochondrial DNA variation. Proteins encoded by mtDNA are involved in each cell's basic energy production, so there is no reason to suspect that they functionally underlie coat-color differences. In other words, mtDNA alleles should be strictly neutral with regard to natural selection for or against particular pelage shades. Furthermore, mitochondrial genes are housed in the cytoplasm of each cell, and hence are physically unlinked from those in the cell's nucleus. Results of the PCM analyses involving mtDNA and pelage colors are shown in Fig. 3.1. They indicate that at the Pinacate site, light and dark mice are thoroughly intermingled among mini-branches of the mtDNA phylogeny. Thus, no evident population-genetic structure exists at this location for selectively neutral genes, a finding that adds further support for the conclusion that allelic variation at the MC1R gene is indeed functionally responsible for the coat-color polymorphism in pocket mice at this Arizona location.

An ongoing challenge for evolutionary biology is to identify functional connections between molecular genotypes and organismal phenotypes for fitness-related traits. The task is difficult because each case requires a fruitful conjunction between at least two lines of hard-to-come-by evidence. First, phenotypic traits must be identified that are ecologically important and can be shown to affect genetic fitness in particular environments; and second, specific genes must be identified and documented at the molecular level to causally underlie those phenotypic adaptations. Coat-color variation in pocket mice is one of relatively few examples in which both of these criteria have been met, and PCM analyses played an important role.

### Sexual dichromatism

In many avian species, one sex (usually the male) is more brightly colored than the other. For example, male Red-winged Blackbirds (*Agelaius phoeniceus*) are solid black with bright red shoulder patches whereas females are brown and streaked (see drawing); and Summer Tanagers (*Piranga rubra*) are brick red whereas females are

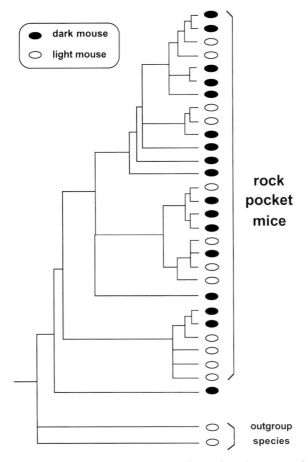

*Figure 3.1.* mtDNA micro-phylogeny for Rock Pocket Mice at the Pinacate site in
Arizona (after Nachman *et al.*, 2003). Note that light-pelage and dark-pelage mice
(which are strongly associated with particular habitat types) are intermingled along
branches of this gene tree, indicating a lack of appreciable population-genetic
substructure in this selectively neutral trait despite pronounced population-genetic
structure with respect to adaptive pelage colors.

drab olive-yellow. In other species, male and female plumages are nearly identical,
being either uniformly dull as in the House Wren (*Troglodytes aedon*), or bright
as in the Blue Jay (*Cyanocitta cristata*). Some bird species even show geographic
variation in degrees of sexual dichromatism. Plumage dichromatism is merely one
aspect of sexual dimorphism, which also includes any differences between the two
genders in other sexual adornments (such as the pronounced red wattles on the
head of a male turkey), or in body size.

In a proximate, mechanistic sense, sexual dichromatism results from sex-biased
gene expression. Males and females of a given species typically share the same genes

Red-winged Blackbird pair

(except for a few loci on sex-specific chromosomes), but their phenotypes may differ owing to distinct developmental programs, often mediated by gender-related hormone profiles. In some avian species, bright colors develop only in the presence of testosterone, whereas in others estrogen concentrations are the key, inducing dull and bright plumages when low or high, respectively. In one recent PCM exercise, Kimball and Ligon (1999) concluded that estrogen-dependent dichromatism is probably the ancestral condition in birds overall, and that testosterone-dependent dichromatism is derived.

In an ultimate, evolutionary sense, various selective pressures in the physical and social environment presumably have influenced where a given species falls on the monochromatism–dichromatism scale. Traditionally, the magnitude and diversity of sexual dichromatism in birds were assumed to be due to variation in the intensities of sexual selection on male plumage. Like natural selection, sexual selection entails differential reproductive success, but with the differing fitnesses in this case due strictly to variation among individuals in their ability to acquire mates. Sexual selection may be intrasexual (e.g. fighting among males for territories) or intersexual (e.g. mating preferences of females for particular males). Thus, a common supposition was that when cocks are more colorful than hens it is because brighter males of that species tend to fare better in mate acquisition, either via success in male–male competition or more directly by being physically attractive to mate-prospecting females.

However, contemporary sexual selection on males, by itself, is unlikely to explain where all avian species fall along the sexual dichromatism–monochromatism scale. First, natural selection can in many cases counter or even override sexual selection (as when predation rates on bright males are unduly high). Colorful males might fare well in terms of mating opportunities, but none the less be reproductively disadvantaged on average by being more susceptible to predators. Second, evolutionary transitions between dichromatism and monochromatism may be due to the gain or loss of colors or adornments in females as well as males. Indeed, at least one comparative study concluded that evolutionary alterations in female plumage have accounted for switches between avian dichromatism and monochromatism just as often as have changes in male plumage (Peterson, 1996). Finally, the current state of plumage chromatism in any taxon is likely a partial reflection of historical legacies as well as contemporary selective forces, meaning that phylogenetic considerations are important too. These can be addressed via PCM exercises, which have generally confirmed, however, that patterns of sexual chromatism are evolutionarily highly labile in birds.

Recent PCM analyses have provided other insights into the nature of avian plumage evolution. For example, several studies reviewed by Wiens (2001) indicate that evolutionary changes in sexual ornamentation often precede rather than postdate changes in mating preferences, thus implying that non-selective factors (such as genetic drift) may have been involved in various instances. Another surprising conclusion from PCM analyses is that losses of elaborate male plumages have actually been much more common than their gains in several clades. In tanagers (Thraupidae), for example, a PCM investigation of 47 genera revealed that the probable ancestral condition was dichromatism with colorful males, and that subsequent evolutionary transitions from bright to dull color were about five times more likely than changes in the opposite direction (Burns, 1998). Similarly, in pheasants (Phasianidae) the ancestral condition was inferred to have been sexual dimorphism from which, in the most derived species, less-ornamented males eventually evolved (Kimball *et al.*, 2001). And in a PCM analysis of almost

5300 species of passerine (perching birds), Price and Birch (1996) found that evolutionary transitions between sexual chromatic patterns had occurred *at least* 150 times, with the mean probability per lineage of a switch from dimorphism to monomorphism being about 2–4 times higher than the probability of change in the opposite direction.

Scientists are still a long way from fully understanding the interacting factors that can shape plumage color variations between the sexes. The important point for current purposes is that PCM findings have challenged the conventional wisdom that current sexual selection acting on males is the primary force promoting avian dichromatism. These PCM analyses have suggested instead the following: that phylogenetic shifts in sexual dimorphism often result from changes in female (as well as male) plumages; that sexual dichromatism (not monochromatism) is often the ancestral (rather than derived) state for particular avian clades; that the contemporary expression of sexual dichromatism often seems to have been the consequence of selection for ornament loss (rather than ornament elaboration); and that such selection for ornament loss can take place in either sex.

A hallmark of cutting-edge science is that it often raises as many questions as it answers. By this criterion, the recent effort to incorporate PCM into studies of avian plumages and sexual selection is indeed pioneering. Although some of these PCM findings remain provisional and controversial, they have certainly enlivened the field.

## Dabbling into duck plumages

Members of the genus *Anas* are known as "dabbling" ducks owing to their habit of splashing about in shallow water while feeding mostly on aquatic vegetation. Rather than fully submerge themselves as do diving ducks, dabblers exhibit a tip-up behavior (see drawing) in which their legs paddle vigorously while their buoyant rear-ends remain above water and their necks stretch down to reach aquatic plants at shallow depths.

In many species of dabbling duck, the drakes during the mating season are much more brightly colored than the hens. Male Mallards (*Anas platyrhynchos*), for example, assume a green iridescent head, yellow bill, and chestnut breast, whereas females retain a uniformly streaked-brown plumage throughout the year. Hens of most *Anas* species are similarly dull and cryptic, whereas breeding drakes develop colorful plumages ranging from bright cinnamon bodies (*A. cyanoptera*), to chestnut heads contrasting with snow-white breasts (*A. acuta*), to harlequin mixes of diverse colors on the head and body (e.g. *A. formosa*). In some dabbling ducks

Northern Pintail Ducks

(e.g. *A. fulvigula*), the drakes resemble the dull-colored hens year-round. These species are termed monochromatic rather than dichromatic.

The prevalence of colorful drakes and cryptic hens in the genus *Anas* contributed to the conventional wisdom that sexual dichromatism was probably the ancestral condition for dabbling ducks, from which monochromatism evolved secondarily and independently on various occasions (such as in *A. fulvigula*). This thesis appeared to be challenged from initial PCM analyses, but upon further inspection all bets based solely on PCM are now off. The primary purpose of this section is to show how phylogenetic inferences can be highly dependent upon the particular model of evolution assumed for the character states in question. It is a cautionary tale that applies in principle to most PCM studies (see Appendix), but especially those that deal with evolutionarily labile phenotypes such as plumage colorations in birds (see also the previous essay).

As described in the Appendix, parsimony has been the most widely adopted criterion for deciding between alternative hypotheses about patterns of character-state evolution. In the context of PCM, parsimony is usually interpreted to mean that phylogenetic reconstructions involving fewer character-state changes during evolution are preferable to those requiring many more character-state transitions to account for the observed distributions of phenotypes in extant species. The basic idea of parsimony is that simple evolutionary explanations are generally to be preferred over more complicated ones. However, as this example will show, notions of what are simple and complex in evolution can themselves be very complex.

Figure 3.2 shows a molecular phylogenetic tree for nearly 50 *Anas* species, and also plots the current distribution of sexual dichromatism versus monochromatism in those extant species. The left-hand version of the diagram shows, for various nodes inside the tree, ancestral character states as inferred from a computer-based maximum-parsimony analysis under the assumption that evolutionary transitions between dichromatism and monochromatism were equally likely in either direction. At face value, this PCM reconstruction implies that plumage monochromatism was the ancestral condition at most of the intermediate and deep nodes in the *Anas* phylogeny. Thus, these results appeared to contradict conventional wisdom that plumage dichromatism was the ancestral state for puddle ducks and that monochromatism evolved independently on several occasions within *Anas*.

However, the right-hand version of Fig. 3.2 shows a very different PCM interpretation of these same data, this time based on a maximum-parsimony analysis in which it was assumed *a priori* that evolutionary losses of plumage dichromatism are five times more likely than evolutionary gains. Under this model, all intermediate and deep nodes in the phylogeny show dichromatism as the probable ancestral state, implying that monochromatism in various groups of extant species is a derived condition with multiple origins. In other words, this second PCM reconstruction (based on a different interpretation of what might be most parsimonious) is entirely consistent with traditional ornithological wisdom concerning the evolutionary history of alternative plumage conditions in puddle ducks. It is also consistent with findings from several other avian groups that monochromatism probably evolved from ancestral dichromatism on multiple occasions (see previous section).

The scientist – Kevin Omland – who conducted these alternative PCM analyses on puddle ducks went on to argue (Omland, 1997) that the second evolutionary model described above is probably more plausible for a variety of biological reasons (independent of phylogenetic considerations *per se*). He suggested, for example, that a presumably complex phenotype such as plumage dichromatism is mechanistically more likely to be lost than gained independently during the evolutionary process. Also, closer inspection revealed several instances (not accounted for by the species-level phylogeny shown in Fig. 3.2) in which monochromatic groups of puddle ducks were phylogenetically nested within more widespread dichromatic species, again suggesting that sexual dichromatism preceded monochromatism at these micro-phylogenetic levels.

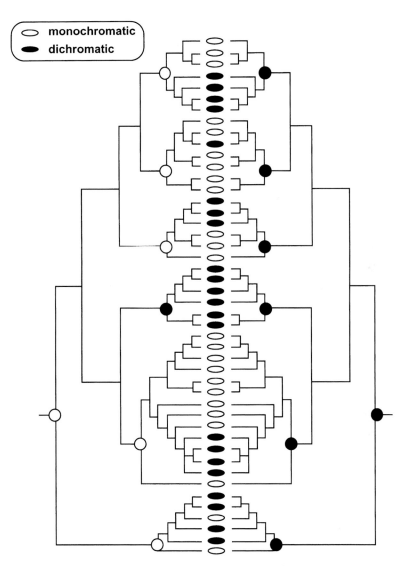

*Figure 3.2.* Two alternative phylogenetic reconstructions of ancestral plumage conditions in puddle ducks (after Omland, 1997). Left, parsimonious outcome based on a model in which it was assumed *a priori* that evolutionary transitions between plumage monochromatism and dichromatism are equally likely in either direction; right, parsimonious outcome based on a model in which it was assumed that evolutionary losses of sexual dichromatism are five times more likely than gains. Ovals indicate plumage conditions of different living species of puddle duck; circles indicate inferred states for particular (not all) interior nodes of the tree.

Regardless of which evolutionary scenario (or some combination between the two) in Fig. 3.2 is correct, the key point here is that PCM reconstructions can sometimes be highly sensitive to subtle variations in the underlying evolutionary assumptions. Indeed, "sensitivity analysis" is the term given to PCM approaches where alternative evolutionary models are purposefully explored in order to assess the robustness or conclusiveness of particular historical reconstructions.

### Specific avian color motifs

The world's approximately 10 000 living species of bird display tremendous plumage variety, but also some recurring themes of body coloration and pattern. No fewer than 26 species possess a black cap that is sufficiently prominent as to have been included in the species' common name (Black-capped Chickadee, Black-capped Kingfisher, Black-capped Gnatcatcher, Black-capped Petrel, etc.). Other such examples include 41 species (in 40 different genera) with "black-throated" in the name (such as Black-throated Blue Warbler), 29 with "red-billed" (e.g. Red-billed Toucan), and nine with "orange-breasted" (Orange-breasted Sunbird). Species sharing any such epithet often belong to different taxonomic families or orders, so each descriptive plumage feature of this sort clearly has arisen on multiple independent occasions during avian evolution.

Sometimes even closely related species have evolved particular recurrent plumage motifs. For example, in PCM analyses based on mitochondrial and nuclear phylogenies of 44 different patches of feather color in numerous species of New World oriole (genus *Icterus*), Omland and Lanyon (2000) deduced that 42 of those plumage features (95%) had experienced multiple evolutionary gains or losses within this taxonomic group. As an illustration, "black pigmentation on the lesser wing coverts" apparently evolved on at least half a dozen separate occasions in these orioles (see Fig. 3.3). Furthermore, striking examples were documented in which entire suites of shared plumage patterns, such as white wingbars plus black throat, black back, and black tail, originated more than once, independently, in various *Icterus* lineages.

Avian plumages presumably evolve under joint influences of natural and sexual selection, but the precise ecological or evolutionary factors promoting recurrences of specific plumage motifs have remained unclear. Two general classes of possibility are sometimes distinguished. First, perhaps a nearly unlimited palette of achievable plumage patterns exists in principle, but recurrent selective agents (such as lighting conditions in the bird's environment, or female preferences for specific patches of body color on males) confine evolutionary outcomes to the subset of alternative motifs actually observed. Alternatively, perhaps genetic or developmental

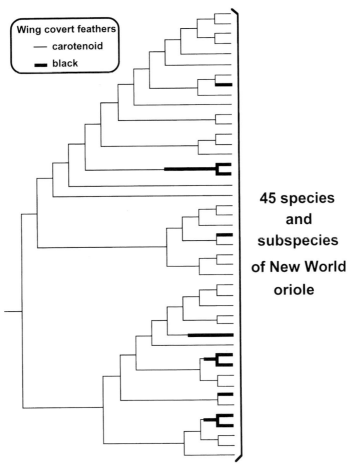

*Figure 3.3.* Molecular phylogeny (estimated from mtDNA sequences) for 45 species and subspecies of New World oriole (after Omland and Lanyon, 2000).

constraints somehow restrict the number of available plumage options, such that recurrent outcomes reflect repeated evolutionary transitions among a finite number of achievable alternatives. The truth probably lies somewhere between these two hypotheses, but only a compilation of empirical outcomes (including identification of the genetic bases of plumage patterns in various avian taxa) will ultimately decide the issue.

One of the first such investigations was conducted only recently, and it yielded a surprise. Using sophisticated laboratory techniques, Nicholas Mundy and colleagues dissected and compared molecular genetic mechanisms underlying the independent evolution of melanic plumage patterns in the Snow Goose (*Anser caerulescens*) and Arctic Skua (*Stercorarius parasiticus*), a type of gull. Within each of these species, some individuals have mostly white plumage, others are mostly dark (melanic), and others are intermediate in color. At the outset, scientists had little reason to suspect that the distinctive melanic forms in these two species would have similar (much less identical) genetic bases because these taxa are distantly related, and because melanistic patterns in some other animals can be quite complex. More than 100 genes are known, for example, to affect the amounts and distributions of melanin in pelts of laboratory mice. But in these geese and skuas alike, specific amino-acid alterations in a protein produced by the melanocortin-1 receptor or MC1R gene proved to be responsible for interconversions between the white and the melanistic plumages. (Indeed, this same protein also appears to be responsible for light versus dark pelages in some rodents; see *Light and dark mice*, above.)

The MC1R protein resides in melanocytes (specialized pigment-producing cells). There, depending on its precise molecular structure, the protein binds to specific circulating hormones in the animal's body and thereby acts as a powerful on–off switch for melanin production. Melanistic versions of the MC1R gene can be partly dominant to the non-melanistic alleles, so heterozygous birds display plumages that are intermediate between those of the white and dark homozygotes. Furthermore, in both the geese and skuas, dark appears to be the derived plumage condition. By comparing magnitudes of genetic variation in the derived versus ancestral gene sequences, and interpreting results in the context of a molecular clock, Mundy and colleagues estimated that these color polymorphisms within each species may be several hundred thousand years old. Field studies have shown that the white and dark plumages influence mate choices and thus are subjects of strong sexual selection.

The implication from this particular example is that a small number of genetic mechanisms can underlie at least some salient and repeated color transformations in birds (and other animals). The geese and skuas are unrelated, but each has evolved a dramatic intraspecific polymorphism of light versus dark plumages for which comparable mutations in the same gene proved responsible. This raises a broader question: might various other recurrent color patterns in birds similarly be governed by key evolutionary alterations involving a limited number of "genetic cassettes," or are the mechanistic controls far more diverse and species-idiosyncratic? In other words, do any genetic commonalities underlie, for example, the features that at least superficially are shared by the Purple-throated Sunbird and Purple-throated Cotinga, Rufous-vented Warbler and Rufous-vented Tit, Yellow-billed Shrike and Yellow-billed Magpie, or the White-eyed Gull and White-eyed Vireo? Obtaining answers to such questions will require far more research, but at

Hooded Pitohui

least there is no shortage of candidate taxa and body-color patches available for genetic dissection.

## The poisonous Pitohui

One day in 1989, Jack Dumbacher, a graduate student doing doctoral research, was trapping birds in the jungles of New Guinea when his lips and mouth suddenly began to tingle and then go numb for several hours. The incident occurred just after Jack had licked his hand for scratches incurred when he tried to untangle a feisty Hooded Pitohui (*Pitohui dichrous*) from his nets. Ironically, this unpleasant experience for Jack was to become a positive event in his own career and an exciting development for the field of ornithology. Jack had inadvertently stumbled upon the first example known (to Western science) of a poisonous bird.

Biologically synthesized poisons are common in many other kinds of organism, and for obvious reasons: they deter predators. Impressive arsenals of noxious compounds are produced by a litany of microbes, plants, and animals. To cite just a few examples, penicillin is a defensive weapon synthesized naturally by molds (and secondarily commissioned by humans) to kill bacteria; fluoroacetate, the active ingredient in some rat poisons, is a deadly chemical produced by several Australian plants as a defense against herbivorous mammals; and homo-batrachotoxin is a predator-deterrent nerve and muscle poison synthesized in the skin of Latin America's poison-dart frogs (see the following section). Upon close chemical inspection, it turned out that homobatrachotoxin is also the poison that permeates a pitohui's skin and feathers. In all of biology, this is one of the most striking known examples of evolutionary convergence, at the biochemical level, between detailed phenotypic traits shared by creatures otherwise as different as a bird and a frog.

Lest they be eaten by mistake, most toxin-containing species have also evolved signals that openly advertise their noxiousness to predators. Poisonous microbes and plants, for example, often emit disgusting odors that deter potential diners. Poison-dart frogs (Dendrobatidae), of which there are several species, are dressed in resplendent colors that predators have come to regard with utmost suspicion (see the following section). For the pitohui, however, it remains uncertain if or how the bird warns off a predator before it is too late. The homobatrachotoxin itself has no apparent odor, but perhaps the bird emits other rancid chemicals that a predator can detect. Or, perhaps something about the color or texture of pitohui feathers repulses a predator before it takes a lethal bite. Throughout its range, the Hooded Pitohui does display a rather bold pattern of black and brick-red feathers (the so-called type I plumage). However, another poisonous and closely related species, the Variable Pitohui (*P. kirhocephalus*), includes several recognized geographical races that are dressed merely in drab grays and browns (type II plumages).

Ecological and evolutionary relationships between the Hooded and Variable Pitohuis have long puzzled biologists. Both species occur sympatrically (i.e. in the same areas) across much of New Guinea, but the Variable Pitohui differs strikingly in plumage from one geographic region to another in more-or-less checkerboard fashion. At some locales, the Variable Pitohui has a bold black-and-red plumage (type I), just as do all of the Hooded Pitohuis, but at other locations the drab type II plumage prevails. What might account for these odd distributions?

Where these two species closely resemble each other, one evident hypothesis is that both are Müllerian mimics whose bold type I plumages arose through con-vergent evolution for predator repulsion. The phenomenon of Müllerian mimicry, named after a nineteenth-century naturalist (Müller, 1879), is a form of mutual-reinforcement advertising wherein two or more poisonous species have evolved similar warning devices to alert predators to their toxic makeups. Such adapta-tions are particularly well known in butterflies. For example, some South Amer-ican species of *Heliconius* butterflys are unpalatable to birds and reinforce their antipredator messages via geographic covariation in their bright wing-color pat-terns (see *Müllerian mimicry butterflies*, below). So, might the type I plumage

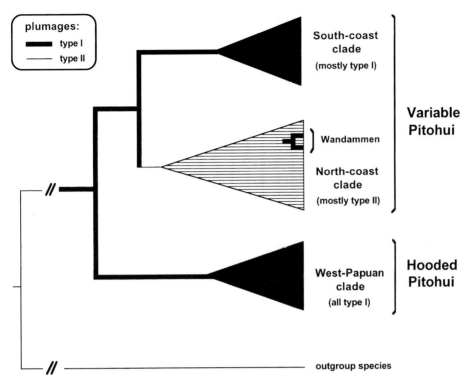

*Figure 3.4.* Molecular phylogeny estimated for various geographic populations of Variable and Hooded Pitohui birds (after Dumbacher and Fleischer, 2001). This depiction shows the type I plumage of the Wandammen population as deeply embedded within the otherwise type II North-coast clade.

shared by pitohuis likewise be due to convergent evolution in warning colorations? If so, this would be one of the very few documented cases of Müllerian mimicry in birds. Alternatively, the close resemblance between type I plumage patterns in the two pitohui species might be merely an historical artifact attributable to shared retention of the primitive (plesiomorphic) pitohui color pattern.

Recent studies by Jack Dumbacher and Rob Fleischer (2001) have helped to decide between these competing hypotheses. By plotting alternative plumage types onto a pitohui molecular phylogeny, these researchers uncovered evidence mostly against (but in one case for) the proposition that interspecific Müllerian mimicry had arisen via convergent evolution in these birds (Fig. 3.4). First, they found that the type I plumage pattern was probably ancestral to this entire taxonomic assemblage. In other words, for most of the type I races of Variable Pitohui, plumage resemblance to the Hooded Pitohui is probably due to retention of the original condition by both species. However, the convergence hypothesis was favored in one specific instance. On the Wandammen Peninsula of New Guinea's north coast, one population of Variable Pitohui with a type I plumage was deeply embedded phylogenetically within an evolutionary clade of that species whose other members display type II plumages. Thus, in this geographical region, it appears quite likely that Variable Pitohuis secondarily gained, by convergent evolution, a plumage pattern that mimics the warning coloration of sympatric Hooded Pitohuis.

Although these phylogenetic findings are consistent with convergent Müllerian mimicry, they do not prove the case even for the Waldammen population. To do that will require additional laboratory and field work that eliminates several competing possibilities: e.g. that the molecular tree itself might be incorrect; that the current reconstruction of ancestral plumage types on the molecular tree might be wrong; that genes for type I plumage might have entered that population via introgressive hybridization with Hooded Pitohuis; or that different pitohui plumage types might somehow reflect varying environmental conditions more so than underlying genes.

Additional research also will be needed to understand why other populations of Variable Pitohuis on the north coast do not also display the type I plumage. Perhaps, for example, not all populations are equally poisonous, and the less toxic forms might benefit more by having cryptic rather than bold plumages. A final point is that, for the bright type I plumages shared by the two pitohui species elsewhere in New Guinea, the overall PCM findings certainly do not eliminate the distinct possibility that Müllerian-mimicry selection has played an important role in maintaining (as opposed to convergently forging) this plumage resemblance. After all, the artistry of natural selection certainly can shape ancestral as well as derived biological clays.

## Warning colorations in poison frogs

Aposematic coloration refers to the presence in particular prey species of bright warning signals advertising dangerous properties to potential predators. These dangers are typically due to noxious or toxic chemicals in the prey's body (see above, *Poisonous pitohuis*). By blatantly displaying conspicuous colors or patterns that predators learn to associate with distastefulness, aposematic creatures tend to avoid attack. Aposematism is found in many invertebrate groups (such as poisonous nudibranch mollusks and various toxic butterflies), but the phenomenon is also known in some birds, fishes, snakes, and amphibians.

One of the most striking vertebrate examples involves dart-poison frogs (in Dendrobatidae). This monophyletic family, with more than 200 diurnal (day-active) species inhabiting Central and South America, includes both aposematic and cryptic species, with the brightly colored forms possessing noxious skin alkaloids, probably of dietary origin. So toxic are these skin compounds that some native Indians, in order to kill small game, load their blowgun darts with poisons rubbed from the tiny amphibians. It is thought that most if not all of the highly toxic dendrobatid frogs are flamboyantly colored, with various species displaying skins of brilliant blue, green, red, orange, or yellow, usually interspersed with patches of jet black. These colors can be stunningly beautiful to human eyes, and have made dart-poison frogs popular as terrarium pets. However, the function of these brilliant colors in nature is to deter predators, who presumably interpret these same patterns as repulsive rather than attractive. Predation is not merely a hypothetical risk for the dart-poison frogs, as evidenced by the fact that non-toxic dendrobatid species possess cryptic colors that blend in with the environment.

Once aposematism is well established in a toxic species, its evolutionary maintenance is quite easy to understand. However, each evolutionary origin of the phenomenon from a cryptic ancestor poses a conundrum. Whenever an aposematic genetic feature first arises in a toxic prey species, its initial frequency is inevitably low. Furthermore, that trait is unlikely to be favored by natural selection initially, because at its inception a conspicuous color would make an aposematic specimen more (not less) subject to attack by predators, most of whom will be naïve to the danger. Another reason to suspect that aposematism arises only rarely and with considerable difficulty comes from the notion that poisonous alkaloid compounds, by virtue of their toxic properties and structural complexity, might be metabolically challenging for a prey species to synthesize and utilize safely. If these speculations are correct – that each evolutionary transition to toxicity and aposematism poses a substantial biological hurdle – then a reasonable hypothesis is that allied species exhibiting this syndrome might typically be monophyletic (rather than polyphyletic). Such had been the conventional wisdom for the dart-poison frogs.

However, recent analyses of molecular data from dozens of dendrobatid species have painted a very different picture (Santos *et al.*, 2003). Bright, poisonous species proved to be scattered across the dendrobatid phylogeny in ways that suggest at least four (and probably five) independent evolutionary origins of the toxin–aposematism syndrome (Fig. 3.5). One large assemblage of more than 20 poisonous species (in the sister genera *Dendrobates* and *Phyllobates*) constitutes a clade, but smaller aposematic lineages appear elsewhere in the phylogenetic tree. Another surprise emerged from this PCM exercise: namely, most of the aposematic toxic clades consist of species that are dietary specialists on ants, termites, or mites, whereas the cryptic non-toxic species have more generalized diets (Fig. 3.5). Evidence was already available that poisonous dendrobatids sequester at least some of their predator-deterrent toxins (such as izidines and pumiliotoxins) from the ants

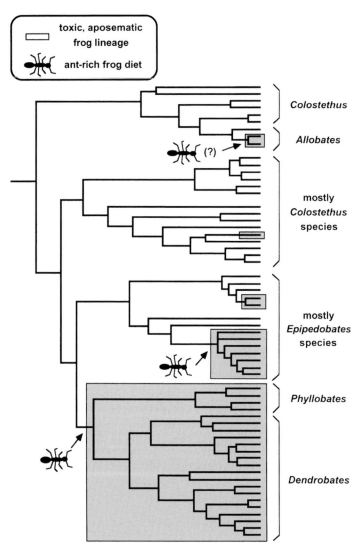

*Figure 3.5.* Molecular phylogeny (estimated from mtDNA sequences) for poison frogs (after Santos *et al.*, 2003).

they eat. Indeed, captive dart-poison frogs reared on fruit flies are known to lose much of their toxicity. The phylogenetic analyses indicate that the evolutionary origins of toxicity and aposematism in dendrobatid frogs are often closely linked to the evolutionary origins of dietary specializations. However, the exact order of phylogenetic transitions from non-toxic to toxic composition, from crypsis to aposematism, and from a generalized to a specialized diet, all remain to be determined through further analyses and experiments. One possibility, for example, is that when frogs happen to acquire higher toxin loads by moving to an ant-rich diet, this sets in motion directional selection pressures for further diet specialization and the evolution of aposematism.

Another phylogenetic study on dart-poison frogs was focused on details of warning colorations in several Peruvian *Dendrobates* species. In the nominal taxonomic species *D. imitator*, different geographic populations had been reported to show three distinctive aposematic patterns: black body spots on a yellow background, horizontal yellow stripes across a black body, and longitudinal yellow stripes down the length of a black body. Each body-color motif in *D. imitator* closely matches that of a sympatric congener (*D. variabilis*, *D. fantasticus*, and *D. ventrimaculatus*, respectively). A molecular phylogeny based on mtDNA sequences confirmed that these diverse forms of *D. imitator* all belong to a monophyletic lineage that is genetically distinct from the other three taxa listed above. Thus, evidently, the aposematic colorations arose on multiple occasions, with sympatric frogs thereby mutually reinforcing their warnings to predators. This study provided the first molecular phylogenetic support, in any amphibian, for a special type of Müllerian mimicry in which one polytypic species mimics different sympatric species in different geographical areas.

## Müllerian mimicry butterflies

One of the most striking known examples of aposematic coloration in the context of Müllerian mimicry (see the previous two sections) involves toxic butterflies of the New World tropics. What sets this situation apart is that the mimetic warnings – wing-color motifs in this case – are exceptionally refined in their details and are often shared by several sympatric species of *Heliconius* and related genera. Perhaps nowhere more so than in these lovely butterflies has the paintbrush of natural selection been wielded more artistically during the coevolution of phenotypic warnings to potential predators.

At least four different mimetic assemblages exist, each containing up to a dozen heliconiine species all sharing a distinctive aposematic wing-color design that advertises the insects' unpalatability to birds. The situation is especially interesting because different "races" of many of these biological species vary dramatically across the continent, and different species often co-vary spatially in their precise wing-color patterns. In other words, sympatric wing-color races of different biological species often resemble one another far more closely than do allopatric races within a species. Some *Heliconius* species contain 20 or more geographic forms, each typically belonging to a sympatric assemblage of two or more species with nearly identical wing-coloration motifs.

Two of the best-studied species are *H. erato* and *H. melpomene*, each of which consists of multiple geographical races that differ dramatically in wing-color designs. For example, one wing-color race in each species has small light dots and long streaks on a black background, another shows large red blotches on the fore-wings, and another models brilliant iridescent streaks on the fore- and hind-wings. Various races in the two species also show striking geographic concordance. For example, the range inhabited by the red-blotch race in *H. erato* is matched closely by the area occupied by a red-blotch race in *H. melpomene*, and so on for approximately 20 other racial pairs in the two species. Thus, at any geographic locale, members of the two species mutually reinforce their anti-predator warnings.

To estimate the general time frame over which such evolutionary convergences in mimetic patterns emerged, Brower (1994) first surveyed mtDNA sequences in wing-color races of *H. erato*. Phylogenetic analyses of these data revealed a basal historical split between all populations east versus west of the Andes Mountains (Fig. 3.6). This molecular finding seems entirely plausible because the Andes must long have been a major geographic obstacle to these animals' movements and hence to cross-mountain gene exchange in this lowland species. The magnitude of mtDNA genetic divergence indicates that the eastern and western assemblages separated approximately 1.5–2.0 million years ago. By contrast, different populations of *H. erato* within either the eastern or western regions proved to be nearly indistinguishable and hence, by this same molecular genetic yardstick, must have had quite recent historical connections. The relative genetic homogeneity of populations within either the eastern or the western regions contrasts strikingly with the geographic heterogeneity in wing-color motifs, and strongly suggests that these radically different wing-color patterns must have emerged rapidly (and also perhaps recurrently) during the evolutionary process (Fig. 3.6).

A follow-up molecular study by Brower (1996) led to qualitatively similar conclusions for *H. melpomene*, albeit with some important additions. In that species, the deepest phylogenetic split distinguished populations in Guiana from those elsewhere in South America. This and further such evidence strongly indicated that the "phylogeographic" histories (see Chapter 7) of *H. melpomene* and *H. erato* were quite different, despite the fact that present-day geographic distributions of their mimetic wing-color races are closely similar. Clearly, specific wing-color phenotypes have evolved convergently between these two species on multiple independent occasions.

Brower's studies provide a fine example of how an apparent evolutionary disagreement between molecular and phenotypic data can sometimes be resolved through careful phylogenetic detective work. In this case, a striking contrast between the geographic distributions of molecular lineages and wing-color phenotypes reflects the likelihood that different evolutionary forces have been at work on

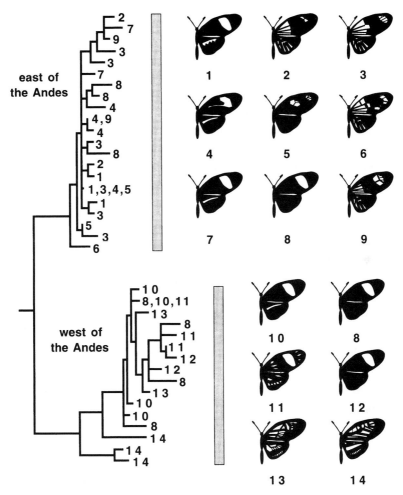

*Figure 3.6.* Intraspecific phylogeny (estimated from mtDNA sequences) for different geographic populations and wing-color races of *Heliconius erato* across tropical South America (after Brower, 1994).

these two sets of characters. In other words, with regard to the evolution of mimetic wings, Müllerian selection pressures must have been strong enough to overwrite the signatures of phylogeny that still seem so legible in presumably neutral molecular markers such as mtDNA.

## Caterpillar colors and cryptic species

Extrapolated estimates of the number of living species on earth range from just a few million to more than 100 million. However, at present only about two million of these actually are known to science, and formally named. Many of the remainder are undescribed species suspected to exist in poorly explored realms (such as the deep sea, and rainforest canopies) or in poorly studied taxa (such as many microbial and invertebrate groups). One reason for supposing that plethoras of species remain to be discovered is that careful systematic investigations of particular geographic regions or taxonomic assemblages routinely uncover "new" (i.e. formerly undetected) species. In recent times, taxonomists collectively have described on average about 18 000 such species per year (Wilson, 1992).

Molecular phylogenetic appraisals have played a key role in the discovery and description of many of these new species. A case in point involves one group of skipper butterflies in the Neotropics. Previously, populations ascribed to a single species, *Astraptes fulgerator*, were thought to be distributed from the southern United States to northern Argentina, and to occupy habitats ranging from deserts to rainforests and from urban gardens to pristine natural areas. Is one species truly so catholic in its tastes, so broad in its adaptive scope? Recent phylogenetic analyses indicate not. Instead, it now seems likely that at least ten cryptic species, each a habitat specialist, had been masquerading as one.

The initial evidence came from studies of the mitochondrial cytochrome oxidase I (COI) gene. By sequencing COI from hundreds of specimens of *A. fulgerator*, Hebert *et al.* (2004) uncovered deep and previously unsuspected genetic subdivisions (Fig. 3.7), of a magnitude normally characteristic of distinct species in many other taxonomic groups. These molecular phylogenetic findings strongly suggested (but by themselves could not prove) that multiple biological species, reproductively isolated from one another, were represented in the collections.

Further inspection revealed that these phylogenetic subdivisions correlated perfectly with previously overlooked differences in these butterflies' natural histories and color patterns. For example, caterpillars in one of the skipper clades ("trigo" in Fig. 3.7) have narrow yellow rings encircling their black bodies, and they feed exclusively on plants in the genus *Trigonia*, whereas caterpillars in the "lohamp" clade (Fig. 3.7) have bold yellow dots along their flanks and specialize on food plants in the genus *Hampea*. Several skipper types were distinguished by other habitat preferences as well. For example, lohamp butterflies inhabit lowland rainforests primarily, whereas hihamps specialize on middle elevation cloud forests and trigos extend into dry forests. Notwithstanding these habitat proclivities, another key discovery was that several of the butterfly clades have overlapping ranges to various extents. All of these field observations, taken in conjunction with the molecular phylogenetic evidence, confirmed the presence of at least ten distinct and often partly sympatric gene pools within "*A. fulgerator*." In other words, multiple biological species are genuinely present within this assemblage of skipper butterflies.

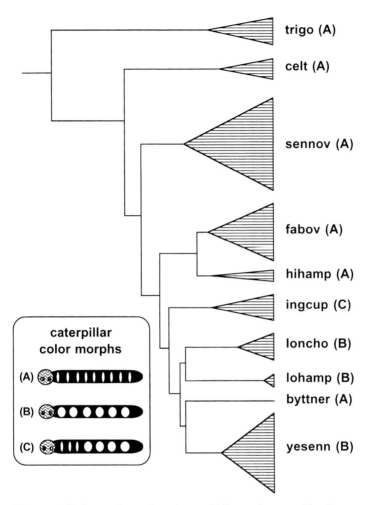

*Figure 3.7.* Phylogenetic tree, based on mtDNA cytochrome oxidase I sequences, for more than 450 individuals in the *A. fulgerator* complex of skipper butterflies (after Hebert *et al.*, 2004). Distinctive clades (deemed to be cryptic species) are indicated by triangles and abbreviated lowercase names. Upper case letters indicate different caterpillar color patterns (shown as caricatures in the inset).

This study provides a fine example of how molecular phylogenetic findings can prompt systematic re-evaluations, resulting in the recognition of "new" species (see also *Coral conservation* in Chapter 7). Of course, outcomes can also be precisely the reverse if (as sometimes happens) two or more recognized taxonomic species prove upon closer inspection to be a single undifferentiated gene pool (i.e. one biological species). Either way, it should be clear that molecular phylogenetic appraisals can play key roles in properly describing the planet's biotic diversity.

In recognition of this fact, a bold scientific initiative known as "DNA barcoding" recently got underway. The basic idea, illustrated by the study on skipper butterflies, is to re-examine all sorts of organisms by using standardized and well-characterized molecular "yardsticks" (such as the COI gene) to help identify salient phylogenetic partitions, and thereby improve our understanding of species boundaries and bio-diversity patterns.

Even as thousands of new species are being described each year, many thousands of others are becoming extinct, mostly as a result of human environmental impacts. Does this mean that global biodiversity might remain fairly constant, on balance? No. Biodiversity is what it is, regardless of whether it is described adequately in our taxonomic summaries. Each discovery of a "new" species is an artificial or anthropocentric event, entailing no change whatsoever in the underlying biological reality. Each extinction event, by contrast, is biologically real, and irrevocable. Thus, even if our net tallies of recognized species should happen to increase dramatically via numerous discoveries of formerly cryptic entities in nature, the sad reality is that extant global biodiversity remains in sharp decline.

# 4 | Sexual features and reproductive lifestyles

Nowhere is the biological world more interesting, flamboyant, and sometimes devious than it is with regard to sexual traits and reproductive activities. Procreation
is the name of the evolutionary game, and organisms have discovered seemingly
endless ways to transmit copies of their genes successfully to subsequent generations. Thus, species display an amazing variety of procreative adaptations and
reproductive lifestyles whose geneses and historical interconversions are of special interest. Case studies highlighted in this chapter illustrate how PCM analyses
have contributed to scientific understandings of the evolutionary transformations
involving a variety of organismal characteristics directly related to reproduction,
such as parthenogenesis (virgin birth), male pregnancy, egg laying, egg dumping, live bearing, delayed implantation, nest construction, brood care, and foster
parentage. The PCM approach has even provided one definitive answer to the
proverbial question: which came first, the chicken or the egg?

## The chicken or the egg?

A chicken's egg is a rather amazing reproductive contraption: tapered at one end
so that it can't easily roll out of the nest; just large enough to provision an embryo
with adequate nutrients yet small enough to pass through a hen's cloaca (vaginal
opening); and covered with a protective calcareous shell that can withstand the
weight of an incubating parent but none the less is fragile enough to permit an
emerging chick to peck its way free. Inside an egg's hard casing are all the biological paraphernalia necessary to support and nourish a developing embryo. These
include a rich yolk that cradles the tiny bird like a catcher's mitt and provides it
with nutrition; a shock-absorbing layer of albumen that also supplies the embryo
with water; an allantoic sac that helps the embryo respire and serves as a septic
tank for safely storing the embryo's poisonous nitrogenous wastes; and several
specialized membranes that surround and separate all of the above. In short, as
a self-contained out-of-body incubator, it's hard to imagine a better evolutionary
contrivance than the egg.

Chickens and egg

The genesis of an egg begins deep within a hen (or, alternatively, the genesis of a hen lies deep within an egg). Let's begin with the hen. At the upper reaches of her reproductive tract is an ovary (most birds have only one, located on the left side) filled with oocytes or ova (unfertilized egg cells) at various stages of maturation. As each ovum matures in its ovarian follicle or pit, it swells greatly in size and acquires a surrounding yolk. On a more-or-less daily ovulation cycle during the breeding season, one mature ovum is released into the upper region of the oviduct, known as the infundibulum. There, fertilization may take place if the hen has recently mated. The fertilized egg (henceforth an embryo as the cell begins dividing and multiplying), with yolk, then starts to move down the oviduct. It first encounters a region known as the magnum, where it settles for about an hour while albumen is laid down. The embryo then shifts to the isthmus region of the oviduct where it resides for an hour or two while various egg membranes are added. The third section of the oviduct, known as the uterus, is where the embryo and its associated structures become encased in a calcium carbonate shell, a process that takes about 20 hours. Finally, the hen pushes the familiar, finished egg out of her cloaca.

In terms of reproduction, eggs are a hen's way of making more chickens (or perhaps chickens are an egg's way of making more eggs). So, which came first? With

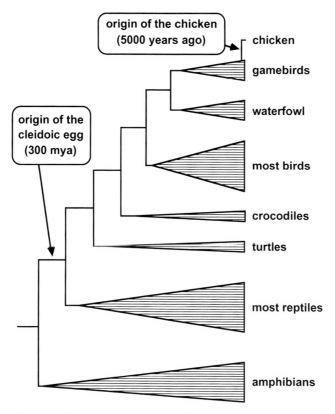

*Figure 4.1.* Phylogeny for representative vertebrates, showing the approximate evolutionary origins of domestic chickens and cleidoic eggs.

respect to an individual's development or ontogeny, this question has no answer, because eggs and hens are merely alternating phases of a continuous chicken lifecycle. However, in a phylogenetic sense, the answer is clear and unambiguous: eggs came long before chickens (Fig. 4.1).

The domesticated chicken (*Gallus domesticus*) is a recent descendent of the Red Junglefowl (*Gallus gallus*) of Southeast Asia. Chickens were domesticated about 5000 years ago, probably in India, and they are also known to have been present in China by 1500 BC and in Greece by 700 BC. Chickens and Red Junglefowl are merely two of more than 200 extant species of gamebird in the taxonomic order Galliformes, which also includes turkeys, partridges, pheasants, francolins, guineafowl, grouse, and quail. From molecular genetic evidence, the closest modern relatives of Galliformes are waterfowl (Anseriformes), with about 160 species. Early in the avian evolutionary tree, the galliform–anseriform clade separated from other basal evolutionary lineages that eventually led to the remaining 20+ taxonomic orders (and approximately 10 000 living species) of birds.

Chicken eggs, like those of all other birds, are closed or "cleidoic." Also possessing cleidoic eggs are reptiles: crocodiles and alligators (the closest *extant* relatives of birds), turtles, most snakes and lizards (although about 20% of these species give birth after the offspring hatch inside the body), and probably most extinct reptiles including the small dinosaur-like ancestors of birds that lived more than 150 million years ago. Phylogenetically speaking, birds are merely feathered reptiles, one subset of a far more ancient reptilian clade. The cleidoic egg that evolved in primitive reptiles, more than 300 million years ago, was a key innovation that enabled the colonization of land by vertebrate animals. By contrast, most fishes and amphibians (including those in the evolutionary line leading to the first reptiles) produce shell-less, often gelatinous eggs that are entirely dependent upon an aqueous environment for survival and successful development. Thus, the ancient reptiles and their avian descendents, forever released from the requirement of depositing their eggs in water, gained new freedoms to venture farther afield as they colonized the terrestrial and aerial realms.

So, the proverbial question has an unambiguous phylogenetic answer. The (cleidoic) egg came well before the chicken – by about 300 000 000 years.

### The avian nest

When a fertilized egg exits the avian body, its next usual repository is a nest where further development of the embryo, and later of the hatched chick, takes place. Collectively, avian nests display great architectural variety. They can range from simple depressions scraped in sand or gravel by various plovers and other shorebirds, to intricate pendulous baskets fabricated by orioles and weaver finches. The prize for the most accomplished designs may belong to African weavers such as the Red-vented Malimbe, *Malimbus scutatus*, a skilled artisan whose nest has a long tubular foyer or atrium (see drawing). Many birds nest in excavated or adopted cavities such as tree hollows (used, for example, by woodpeckers and parrots) or earthen burrows (used by kingfishers and nocturnal petrels and auklets, among others). Grebes often nest on a floating platform of reeds, and some Australian moundbuilders engineer a compost-heap nest of decaying vegetation and sand. Many other birds fashion cup-like nests from such substances as sticks, grasses, leaves, mud, lichens, spider webs, or even, in the Edible-nest Swiftlet (*Collocalia fuciphaga*), their own hardened saliva. In Southeast Asia, these swiftlets' nests are highly prized as a main ingredient in birds-nest soup.

Although nests are crafted from natural materials in the bird's environment, their design and construction reflect innate behaviors of the animals themselves. Thus, avian nests can be thought of as exogenous phenotypes that are subject to

Red-vented Malimbe with nest

evolutionary change just as are endogenous body parts. Accordingly, considerable scientific attention has been paid to uncovering the evolutionary histories of different nest construction motifs.

One categorization that has been addressed is whether a design is "safe" or "open." Safe nests include burrows, cavities, or other enclosures with secure domes.

They also include nests located in protected sites such as predator-free islands, inaccessible cliff faces, or in dense colonies where adults can provide continual monitoring. Open designs, by contrast, include non-colonial cup nests, ground nests, and other accessible sites, presumably with greater predation risk. By charting the taxonomic distribution of safe versus open nests onto a molecular phylogeny for more than 50 major lineages of birds, ranging from quail and ducks to warblers and sparrows, Owens and Bennett (1995) deduced that independent transitions between safe and open nests had occurred several times over the course of avian evolution. These switches between nesting modes apparently occurred in both directions, i.e. from open nests to closed nests as well as vice versa.

These scientists then examined numerous life-history parameters (such as incubation period, age at first breeding, and lifespan) for possible associations with the directions of change in nest design; they found an especially strong correlation between the evolution of safe nests and evolutionary transitions to low pre-fledging mortality. For example, both egg and nestling survival proved to be significantly higher in avian lineages that were characterized by hole or colony nesting. These empirical findings in turn were interpreted as consistent with a longstanding hypothesis that species with high pre-breeding mortality tend to experience strong natural selection for "fast-lane" life-history traits (such as rapid growth and high fecundity), whereas species with lower mortality rates in the nest more often optimize their reproductive success via a contrasting suite of "life-in-the-slow-lane" features (such as delayed breeding and lower fecundity). Overall, this PCM exercise indicated that variation in the safety of nest design, by virtue of its impact on age-specific mortality schedules, is a key ecological factor affecting how birds in various lineages have evolved distinctive portfolios with regard to how they apportion their investments between survival and reproduction.

Although transitions between different nest architectures clearly occur across evolutionary time, most taxonomic families of bird tend to be fairly conservative with regard to general mode of nest construction. For example, all 30 species of mimic thrush (Mimidae) build bulky cup-shaped nests. A few avian families, however, are anything but conservative in nest design, a good example being the swallows (Hirundinidae). Within this phylogenetic group of 80 living species can be found cavity adopters, burrow diggers, and mud-nest constructors; and among the latter species are those that build an open mud cup, or a partly covered cup, or a fully roofed cup complete with protective eaves. No other songbird family shows such wide variety of nest designs.

To explore the evolutionary history of these alternative housing arrangements, Winkler and Sheldon (1993) estimated a molecular phylogeny of 17 swallow species, onto which they plotted the various nest types (Fig. 4.2). Results indicated the

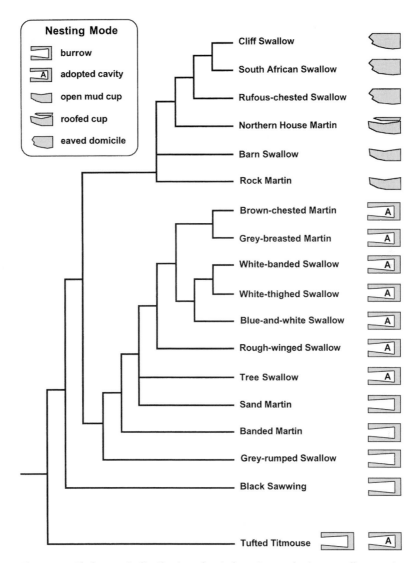

*Figure 4.2.* Phylogenetic distribution of varied nesting modes in 17 swallow species (Hirundinidae) plus an outgroup (after Winkler and Sheldon, 1993).

following: related species in well-defined clades usually shared similar or identical nest designs; the ancestral nesting condition for swallows was probably soil burrowing; and from this inferred ancestral condition, cavity adoption evolved in one major clade (most likely originating in New World forests), and mud-nest construction evolved in another (likely originating in African savannahs). Within the mud-nest clade, the PCM analysis indicated that simple open cups probably came first, followed by an evolutionary progression to roofed and eventually to eaved domiciles in some of the derived species. This study has become a classic example of how living creatures, by manipulating their natural environments, can in some cases drive the "evolution" of complex inanimate structures.

### Egg dumping and foster parentage

Normally, eggs and hatchlings within an avian nest are the biological offspring of the tending parents, but this is not invariably true. Every once in a while, foster children are present. This can happen whenever a female surreptitiously lays one or more eggs into the nest of another bird and then leaves the parenting duties to the (duped) foster parent(s). This phenomenon is known as egg dumping, or, more formally, brood parasitism (BP). The word parasitism comes from the fact that an egg dumper freeloads or appropriates the parenting efforts of other birds towards rearing her own genetic offspring. As in any host–parasite relationship, one interested party (the egg dumper in this case) profits in the currency of genetic fitness at the expense of another party (here, the foster parents).

BP can occur both within and between species. At the intraspecific level, the phenomenon is known to occur sporadically in grebes (Podicipediformes), gamebirds (Galliformes), waterfowl (Anseriformes), pigeons and doves (Columbiformes), and various songbirds (Passeriformes); it tends to be especially common when population densities are high or when nest sites are in short supply. In a few extreme cases, as many as 10%–20% of nests in a local population have been found to contain one or more eggs dumped by nest-parasitic females.

At the interspecific level, BP may be either facultative or obligate, depending on the species. An example of facultative interspecific BP involves Yellow-billed Cuckoos (*Coccyzus americanus*) and Black-billed Cuckoos (*C. erythrophthalmus*) of North America, which occasionally dump eggs into each other's nests. Obligatory egg dumping (undoubtedly the most intriguing form of interspecific BP) is practiced by members of about half a dozen distinct avian groups including New World cowbirds (in Icteridae), Old World cuckoos (in Cuculidae), South American Black-headed Ducks (*Heteronetta atricapilla*; Anatidae), African honeyguides (Indicatoridae), and African whydahs, indigobirds, and parasitic weavers (all in Ploceidae). The scattered distribution of interspecific BP across the phylogenetic tree of birds, and the near certainty that it is a derived condition, are *prima facie* evidence that the phenomenon is polyphyletic, having arisen multiple times independently in different avian lineages. More focused molecular phylogenetic analyses of particular taxa have revealed that interspecific BP is also polyphyletic even within the cuckoos (Aragon *et al.*, 1999) but is probably monophyletic within the indigobird–whydah clade (Sorenson *et al.*, 2003).

Plausible evolutionary pathways to obligate interspecific BP are not hard to imagine. The easy first step probably occurs when a female lays an egg into a nearby nest of familiar design (i.e. of a conspecific). Being of the same species, the nestling might escape the tender's notice as being a stepchild as opposed to a biological offspring. If the foster child is hatched and fledged, the brood parasite will have been genetically rewarded for her opprobrious behavior. Thus, all else being equal, any genes that might predispose an individual for egg dumping would tend to increase in frequency under natural selection. The second evolutionary step, illustrated by the North American cuckoos, probably entails occasional egg dumping into nests of closely related species, where again the foster eggs and hatchlings are likely to resemble the native progeny in appearance and behavior, and thereby escape detection by the foster adults. The third and final evolutionary step is the refinement and elaboration of such facultative BP behaviors into full-blown, obligatory interspecific BP. At each step along that evolutionary progression, the brood parasite is inevitably under strong selection to dupe or otherwise persuade the resident nest tenders to rear foster kids. But throughout that same honing process, members of the foster species will also be under strong selection to avoid being brood parasitized (presuming, as is almost certainly true, that rearing foster children diverts precious parental time and resources that otherwise could be spent in rearing biological progeny).

Thus, as with standard parasites (such as ticks or tapeworms) and their hosts, brood parasites and the species they parasitize are continually engaged in co-evolutionary battles. Behavioral and morphological tactics that have evolved in one or another species of interspecific brood parasite can be amazingly devious: sneaky or skulking behavior by an egg-dumping female in her search for a nest to parasitize; a protrusible cloaca that a BP female can extend into a host's nest that otherwise might be difficult for her to access; a stereotypical behavior in which a female removes or eats an egg from a host's clutch even as she lays her own (as if the regular tenders might count eggs); egg color patterns that match those of the host's eggs (so the foreign egg is less likely to be rejected by step-parents); in foster nestlings, mouth-gape color patterns and begging behaviors that mimic those of the host nestlings and elicit a feeding response from tending adults; ruthless behavior in which a parasitic nestling pushes a host egg out of the nest or otherwise kills host nestlings; and exaggerated food-begging calls and behaviors by foster hatchlings or even fledged chicks. These food-begging tactics are so effective (and the parental instinct to feed offspring is so pronounced) that it is not uncommon to observe, for example, a tiny adult warbler shoving food into the gaping mouth of a several-fold-larger cuckoo chick.

Selection pressures imposed by BP also have led to a variety of behavioral countermeasures by hosts. In one or another host species, these may include nest guarding and mobbing behaviors that tend to discourage egg dumping by foreign females; refusal to feed foster hatchlings or chicks; and desertion or reconstruction of a nest in which a parasitic egg or hatchling has been detected by the resident adults (ergo selection pressures on the brood parasite for egg mimicry and hatchling mimicry). In one of the more interesting nest reconstruction modes, some songbirds, upon detecting a foreign egg, physically "roof over" their current nest before laying a new clutch of eggs in a refurbished nest one floor above. Reports exist in which this kind of home remodeling has been repeated several successive times by a pair of frustrated parents, each in response to the appearance of yet another dumped egg.

In any co-evolutionary war between parties with conflicting interests, precise outcomes at any specified point in evolutionary time are difficult to predict. Interactions between a brood parasite species and its host are likely to be in ever-shifting states of dynamic equilibrium that reflect contemporary impacts of opposing selection pressures but also the idiosyncratic genetic legacies unique to each pair of interactants. None the less, phylogenetic backdrops can sometimes help in testing alternative hypotheses about brood parasitism, as the following survey of cowbirds will illustrate.

Cowbirds are native to the Americas, and several species show various specializations for brood parasitism. With respect to numbers of host taxa utilized, the Brown-headed Cowbird (*Molothrus ater*) of North America parasitizes more than 200 species of small land bird, whereas, at the other end of the spectrum, the Screaming Cowbird (*M. rufoaxillarus*) of Argentina parasitizes only the Bay-winged Cowbird (*M. badius*). (The Bay-winged Cowbird itself is a non-parasitic species, although it raises its young in the deserted nests of other species.) Three other cowbird species fall between the Brown-headed and Screaming Cowbird with respect to how many host species they parasitize: the Giant Cowbird (*Scaphidura oryzivora*) utilizes seven host species, the Bronzed Cowbird (*M. aeneus*) uses about 70, and the Shiny Cowbird (*M. bonariensis*) parasitizes more than 170 species.

The number of host species is of special interest for the following reason. Under one evolutionary hypothesis, host-specificity is the ancestral condition, because, plausibly, interspecific BP first arises when a given species begins to lay eggs in the nest of a closely related taxon. If so, host-generalization would then be a derived condition, i.e. a more catholic behavior that evolves later as the brood parasite

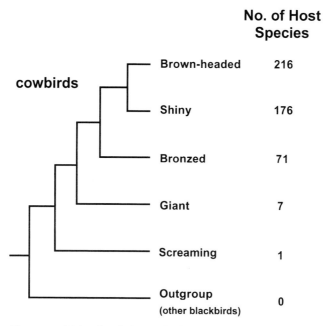

No. of Host
Species

*Figure 4.3.* Molecular phylogeny for brood-parasitic cowbird species, also showing numbers of host species utilized (after Lanyon, 1992).

eventually extends its scope and its BP capabilities. However, precisely the opposite hypothesis – that host-generalization is the primitive condition – also seems plausible, if specialization on a particular host species normally requires an extended period of evolutionary time for genetic refinement. By generating an mtDNA phylogenetic tree for several BP cowbird species, and then superimposing host numbers on that tree, Lanyon (1992) tested these alternative hypotheses, and concluded the following (Fig. 4.3): brood parasitism probably is monophyletic in cowbirds; and host-specificity in this clade is most likely the ancestral condition from which host-generality evolved later in some of the BP cowbird species.

### Egg laying and live bearing

Most reptiles, including all turtles and crocodilians, lay cleidoic (enclosed or shelled) eggs into the environment (see above, *The chicken or the egg?*). They are said to be oviparous. However, about 20% of reptile species (notably various snakes and lizards, in the order Squamata) instead give birth to live young, i.e. they are viviparous. In effect, their eggs hatch inside the body rather than outside, and the pregnant mother subsequently carries and provisions the developing embryos (to varying degrees, depending on the species) until the time of live-birth delivery.

In the earlier scientific literature, it was frequently asserted (without clear documentation) that viviparity in reptiles can evolve readily from oviparity, but not the reverse. The general sentiment seemed to be that any refined production of an eggshell and its associated structures would require special metabolic pathways and female organ systems that would not easily be re-acquired if they were somehow lost during the evolutionary process. However, the evolution of viviparity would also seem to involve the acquisition of complex adaptations, e.g. for fetal respiration and nutrition inside the female's body, for maternal tolerance of alien cells and tissues, and for the delivery process itself. Furthermore, both oviparity and viviparity undoubtedly entail advantages and disadvantages depending upon ecological circumstances, so neither reproductive strategy is likely to be universally superior, and natural selection might be expected to push different species in different directions with respect to egg laying versus live bearing. Thus, on theoretical grounds, evolutionary biologists began to question the assumption that viviparity in various reptile lineages is invariably a derived condition and that oviparity is necessarily ancestral.

Clearly, these issues can only be resolved empirically, a task for which PCM analyses are ideally suited. Indeed, because more than 50 different reptilian groups contain at least some viviparous species, nature has conducted many independent "experiments" that phylogeneticists can examine, by using PCM logic, to identify the directions of evolutionary transition between oviparity and viviparity. Lee and Shine (1998) conducted one pioneering study by mapping the distributions of viviparity and oviparity onto a cladogram generated from combined phylogenetic analyses of morphological and molecular data. Figure 4.4 depicts a representative subset (involving more than 60 species) of this phylogeny, from which it appears that viviparity evolved more than a dozen times, independently, from oviparity, and that few if any successful evolutionary transitions occurred in the opposite direction. This pattern was generally supported in the broader phylogeny as well, which revealed more than 30 probable evolutionary changes from oviparity to viviparity in three major groups of snakes and lizards, but only five or fewer changes from viviparity to oviparity (and all of these were only weakly supported statistically, and thus somewhat ambiguous, by parsimony criteria). Although oviparous reptiles may have evolved from viviparous ancestors on rare occasions, transitions in the opposite direction have been far more common. In other words, viviparity in snakes and lizards appears to have been rather easy to acquire but difficult to lose.

Interestingly, similar PCM analyses have been conducted on elasmobranch fishes (sharks and rays) and polychaete worms. In the fish study, Dulvy and Reynolds (1997) identified about ten origins of viviparity from oviparity, but only two or

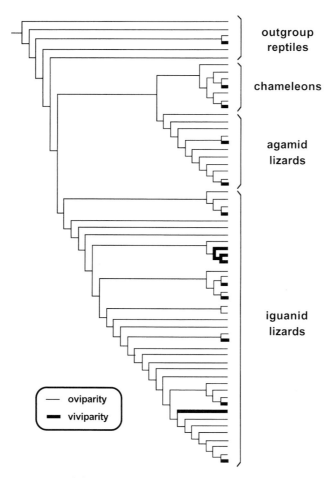

*Figure 4.4.* Phylogenetic tree for more than 60 representative species of lizards and other reptiles, showing multiple evolutionary origins of viviparity (after Lee and Shine, 1998).

three probable origins of oviparity from viviparity. Conversely, in the polychaetes surveyed by Rouse and Fitzhugh (1994), oviparity (in this case "broadcast spawning," i.e. the release of eggs into the open water) apparently evolved from viviparity (larval brooding by females) six times, whereas only one evolutionary transition in the reverse direction was documented.

Several species of snake and lizard are polymorphic for egg laying and live bearing, meaning that some conspecific populations employ one reproductive mode and other populations employ the other. One such species (*Lacerta vivipara*), "caught in the act" of converting between oviparity and viviparity, has been the subject of a detailed PCM analysis. By examining fine-scale mtDNA genealogies of several oviparous and viviparous populations inhabiting different regions in Europe, Surget-Groba *et al.* (2001) concluded that a single evolutionary conversion between reproductive modes had occurred (probably in the eastern portion of this species' range) and that the direction of change had been from oviparity to viviparity. Results are consistent with the trends described above, and they also indicate that such seemingly remarkable evolutionary transitions can occur very rapidly (well within the geological lifespan of a species).

The phylogeographic reconstruction by Surget-Groba and colleagues further suggests that cold climatic conditions during the Pleistocene Ice Ages may have promoted the evolutionary emergence of viviparity in *L. vivipara*, and perhaps in other reptilian lineages as well. According to the cold-climate hypothesis, eggs laid directly into the environment (and the resulting progeny) are at special peril during times of climatic deterioration, when selection pressures might be unusually strong for the within-body retention of fertilized eggs by adult females who can actively seek out suitable microhabitats until their advanced offspring are birthed. Whether this or other ecological factors provided the selective impetus for the evolution of viviparity in snakes and lizards, the PCM analyses indicate that the live-bearing lifestyle, once gained, is extremely difficult for these animals to relinquish.

### Piscine placentas

In most fish species, each reproductively active female discharges hundreds or thousands of eggs into a watery environment where they are fertilized by sperm likewise discharged by spawning males. However, in a small fraction of fish species, females retain eggs within their bodies, receive sperm during mating (males typically possess a modified anal fin or gonopodium that serves as an intromittent organ), carry their developing embryos internally for several weeks, and then give birth to live young. For example, all 130+ species in the live-bearing family Poeciliidae practice this piscine version of pregnancy. Indeed, in some of these fishes, the entire pregnancy process is quite analogous to that experienced by female mammals.

The evolution of internal pregnancy (from egg laying) in any vertebrate group requires that several major biological hurdles be cleared. In general, females must meet the added commitments on time and energy that pregnancy demands, and this typically entails overcoming novel types of challenge such as the suppression of negative immunological interactions between mother and embryos, disposal of embryonic wastes, and proper delivery to embryos of suitable nutrients and respiratory gases. In placental mammals, intermediate stages in the evolution of pregnancy are hard to study directly because the phenomenon apparently emerged more than 100 million years ago and is now well refined in all extant species. However, modern poeciliid fish species vary in the extent to which particular aspects of pregnancy are elaborated, and this makes the group better suited for addressing how (and how often) internal pregnancy evolved.

Of special interest are *Poeciliopsis* fishes. Within this live-bearing genus can be found species ranging from lecithotrophic or yolk-feeding (in which females retain fertilized eggs internally but offer no further nourishment to their developing embryos) to matrotrophic or mother-feeding (in which maternal provisioning of embryos is modest to high). In lecithotrophic species, the mass of an offspring at birth is considerably lower than the mass of an egg at fertilization, owing to the metabolic costs of embryonic development. In the matrotrophic species, by contrast, an offspring at birth may weigh as much as or more than the fertilized egg, implying that its mother had provided nutrients to offset developmental costs. In some cases, an embryo at birth is more than 100 times the mass of an egg at fertilization. Furthermore, among matrotrophic species, levels of maternal provisioning are associated with the extent to which maternal and fetal tissues are elaborated into specialized delivery structures known as placentas (which are absent in lecithotrophic species).

To explore the evolution of maternal provisioning and placental development in *Poeciliopsis* fishes, David Reznick and colleagues used, as historical backdrop, a previously published molecular phylogeny based on mtDNA gene sequences. From this PCM exercise, refined placental structures evidently evolved at least three separate times within *Poeciliopsis* alone (Fig. 4.5), as well as elsewhere in the Poeciliidae. In each case, placental taxa were deeply embedded in clades that otherwise consisted of species with little or no placental development or maternal provisioning of embryos. By examining approximate temporal depths in the phylogeny (based on molecular clock considerations), Reznick *et al.* (2002) further deduced that evolutionary transitions from the absence to the presence of a refined fish placenta can occur in less than 750 000 years.

These researchers also drew possible analogies between the evolution of placentas and that of vertebrate eyes (see Chapter 6, *The eyes have it*). Ever since Charles Darwin expressed amazement at such "organs of extreme perfection," biologists have sought to understand how complex adaptations emerge from far simpler starting conditions. With regard to eyes, Darwin posited that even simple photosensitive organs are probably adaptively advantageous under many ecological circumstances, such that a gradual emergence of refined vision may well have been favored by natural selection at each and every step in the evolutionary progression

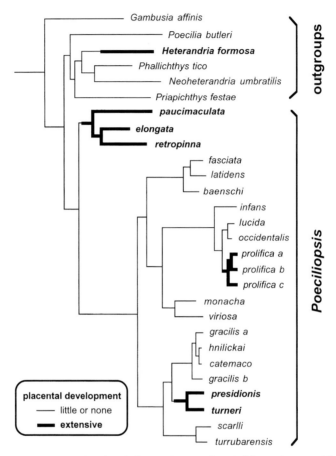

*Figure 4.5.* Molecular phylogeny for *Poeciliopsis* fishes (after Reznick *et al.*, 2002).

to more elaborate visual receptors. Indeed, theoreticians have calculated that highly complex eyes could evolve in as little as about 400 000 years (Nilsson and Pelger, 1994), although this is merely armchair speculation based on plausibility arguments.

Comparable scenarios about rapid placental evolution via intermediate stages might be envisioned, but now with the added weight of empirical evidence. Placentas (like eyes) presumably are complex adaptations whose evolution probably requires cumulative alterations in multiple genes (for example, more than 50 loci are involved in development of the modern mammalian placenta (Rossant and Cross, 2001)). Although no one yet knows precisely how many genes underlie the refined placentas and advanced maternal provisioning of poeciliid clades, the PCM findings do demonstrate that placental evolution in these live-bearing fishes has been both rapid and recurrent.

## Male pregnancy

Quite a different pregnancy phenomenon is displayed by all of the 200+ living species in the family Syngnathidae. In these pipefishes and seahorses, males (rather than females) bear the burden of housing developing embryos. The process begins when a gravid female transfers all or part of her clutch (consisting of dozens to hundreds of eggs) into a specialized brooding area or pouch located under the male's abdomen or tail. The male then fertilizes the clutch with his sperm, and carries the maturing embryos for several weeks before giving birth to offspring that look like miniature versions of the adults. During his pregnancy, the father nourishes, osmoregulates, and protects his brood, whereas the mother plays no role in offspring care.

The male's brooding structure varies greatly among extant syngnathid species. At one end of the spectrum are enclosed and fortified brood pouches with considerable physical complexity, often including placenta-like features. This condition is characteristic of living seahorse species. At the other end of the spectrum are simple and relatively unprotected brooding areas on the male's ventral surface, where eggs are glued but not bodily encased. This condition is found in a few pipefish species. Between these two extremes, various other pipefish species have either thin membranous compartments surrounding each egg, or different types of partly enclosed ventral pouches with protective coverings extending across multiple eggs and embryos.

Anthony Wilson and colleagues (2003) addressed the evolutionary history of syngnathid brood pouches by mapping the phylogenetic distributions of alternative structural designs onto an mtDNA phylogeny for more than 30 representative species (Fig. 4.6). Results generally indicated a good agreement between clade membership and particular types of brood pouch motif. For example, all surveyed members of the *Syngnathus* clade of pipefishes possess closed pouches with two bilateral skin folds grown together; and the *Hippocampus* clade of seahorses was unique in having a full sac-like pouch enclosed by a single covering. On the other hand, each of two or three other pouch designs did recur in different branches of the molecular tree, suggesting a few instances of independent evolutionary origins. Overall, the molecular phylogeny was consistent with the plausible notion that the evolutionary origins of simple brooding designs generally were antecedent to those of brood pouches with greater structural complexity (see Fig. 4.6).

One might suppose that male pregnancy *per se* should qualify syngnathids as being "sex-role reversed" relative to mammals, where females are the pregnant gender. However, in much of the scientific literature sex-role reversal is defined to occur whenever competition for access to mates operates more intensely on

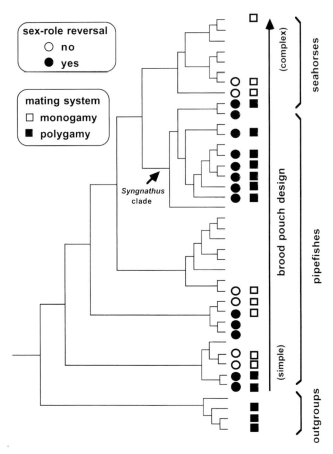

*Figure 4.6.* Phylogenetic tree for 36 syngnathid species and four outgroup taxa, based on mtDNA gene sequences (after Wilson *et al.*, 2003). Mapped onto terminal nodes of the tree are current distributions (where known) of monogamous versus polygamous mating systems and presence versus absence of sex-role reversal among extant species.

females than on males. By this criterion, some syngnathid species are sex-role reversed and others are not. This is evidenced, for example, by the fact that females in some but not all syngnathid species collectively produce far more eggs than can be accommodated within the males' brood pouches, making males a limiting reproductive resource for which females actively compete. Sex-role reversal by this definition has other expected ramifications also: that females should tend to have polygamous tendencies; and that females, more so than males, are likely to display sexually selected phenotypic traits. All of these properties differ diametrically from what is normally observed in species with "conventional" sex roles. There, females tend to be the limiting resource, such that males are under stronger sexual selection and often show elaborate phenotypes for attracting mates (think of peacocks' tails) or for battling amongst themselves for female access (think of rams' horns). A vast scientific literature shows that the topics of sexual behavior, sexual selection, sexual dimorphism (phenotypic differences between males and females), and mating systems can be thoroughly intertwined.

In an effort to sort through some of this complexity, Wilson *et al.* (2003) also mapped empirical findings about sex-role reversal and mating systems onto their syngnathid molecular phylogeny (Fig. 4.6). This enabled the researchers to explore several evolutionary hypotheses. For example, one possibility was that elaborate brood-pouch designs might predict sex-role reversal because males with such pouches might be an even more valuable limiting resource to mate-prospecting females (assuming that more complex pouch designs indicate heavier investments by male syngnathids in offspring care). This expectation was not confirmed, however, by the PCM analyses, which showed instead that sex-role reversed clades were present in both simple-brooding and complex-brooding portions of the syngnathid phylogeny. One possible explanation is that the complexity of brood-pouch design may not be a reliable indicator of relative male investments in offspring care.

Another evolutionary hypothesis, however, gained provisional statistical support from PCM analyses: that sex-role reversal among syngnathid species tends to be phylogenetically associated with polygamous mating by females (i.e. polyandry). For example, all *Syngnathus* pipefishes for which information is available are both polygamous and sex-role reversed (Fig. 4.6), whereas all such species of *Hippocampus* seahorse are monogamous and have conventional sex roles. Furthermore, among pipefishes whose mating systems have been genetically documented to date (using molecular analyses of maternity and paternity [Jones and Avise, 2001]), the species that are more highly polyandrous tend to be those in which females show a greater elaboration of secondary sexual traits (such as bright stripes across the body during the breeding season).

All of the above conclusions are preliminary and warrant further investigation, but a broader point remains. The male-pregnant and sometimes sex-role-reversed pipefishes and seahorses offer researchers a totally fresh perspective on reproductive modes. Indeed, these remarkable little fish have provided a valuable scientific service by demanding that we re-examine many of our traditional mammalian biases about animal mating behaviors.

## Living and reproducing by the sword

In fishes other than the male-pregnant syngnathids (previous section), females tend to be a limiting resource in reproduction, so competition for mates often operates with greater intensity on males. In such species, females can afford to be picky when choosing mating partners, and this (as well as direct male–male competition for mates) can result in intense sexual selection on males. One net effect may be the elaboration of secondary sexual traits that enhance the attractiveness of males to the fairer sex. Probable examples in fishes are the vivid hues of sunfish males (in several species of *Lepomis*) during the breeding season, and the bright body spots and adorned fins of male guppies (*Poecilia reticulata*). Females in these species apparently are attracted to colorful suitors, perhaps merely by feminine whim, or perhaps because male adornments may be honest indicators of high mate quality (such as freedom from parasites or disease). Sexual selection is frequently opposed by natural selection, however. In male guppies, for example, the bright body spots that females find attractive also catch the eyes of predators. Thus, in predator-rich streams in the tropics, natural selection via intense predation has resulted in guppy males that show fewer and less colorful spots than males in predator-free streams, where sexual selection has had freer reign.

Another piscine example of sexual selection on males involves swordtail fishes in the genus *Xiphophorus* (Poeciliidae). As they mature, swordtail males develop a long, colorful extension of the lower portion of their caudal (tail) fin. In old individuals, this rapier-like protrusion may even exceed the length of the remainder of the male's body. The males don't use their swords in combat, but rather during courtship, where they are displayed to curious females. Size does matter in these fishes, as demonstrated by the observation that females, when given a choice, usually prefer to mate with longer-sworded males. Indeed, sexual selection via female preference appears to have been responsible for the whole sworded state of affairs in these fishes.

Swordtails are closely related to platyfishes (also in the genus *Xiphophorus*), the main operational distinction being that swordtail males have sworded tails whereas platyfish males do not. Interestingly, platyfish females also prefer to mate with sword-wielding conspecific males, when given the option. This was demonstrated by laboratory experiments in which researchers surgically grafted swordlike plastic prostheses onto platyfish males, after which female platys consistently chose these "enhanced" males over their normal short-tailed hubbies. These unexpected observations raised a chicken-or-egg-type question: which came first in *Xiphophorus* evolution, male swords, or female preferences for male swords? Given the behavioral observations on female mating preferences described above, clues to the answer might come from phylogenetic analyses. An ancestral position of platyfishes relative to swordtails would be consistent with the female-preference-first hypothesis (also known as the "pre-existing bias" hypothesis), whereas if swordtails

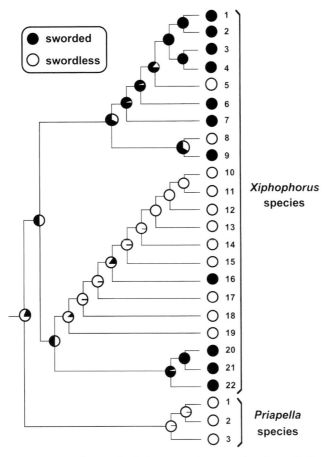

*Figure 4.7.* Evolution of tail phenotypes in 25 species of swordtails, platyfishes, and allies. Pie diagrams indicate the relative likelihoods (from PCM analyses) of specified tail conditions at various nodes in the tree (after Schluter *et al.*, 1997; based on the molecular phylogeny from Meyer *et al.*, 1994).

were ancestral to platys, then swords (but not female preferences for swords) might secondarily have been lost in the derived platyfish clade.

To address these issues, a molecular phylogeny for more than 20 species of *Xiphophorus* was generated and employed as historical backdrop. It turns out that swordless and swordtailed species are intermingled in the *Xiphophorus* phylogeny (Fig. 4.7), with neither assemblage forming a coherent clade. Thus, evolutionary transitions between presence and absence of swords appear to have been rapid and recurrent within the genus, thus making determination of the ancestral state for this clade highly equivocal. However, outlier species in the sister genus *Priapella* consistently lack a sword, so the shared ancestor of the more ancient *Priapella* +

*Xiphophorus* clade was probably swordless (see Fig. 4.7). Interestingly, behavioral experiments have shown that females of at least some *Priapella* species also prefer to mate with conspecific males adorned with long prosthetic tails. Thus, taken altogether, the available data provide substantial (albeit not definitive) evidence for the pre-existing bias hypothesis in which a female preference for sword-wielding males pre-dates the evolutionary appearance of the swords themselves.

The molecular phylogeny also indicates that swords have been lost (e.g. in species 5 and 8 in Fig. 4.7) as well as gained on multiple occasions within the *Xiphophorus* clade. This suggests that despite their attractiveness to females, swords may otherwise be a substantial burden for males. Perhaps they are energetically expensive to produce and maintain, or perhaps they are cumbersome and compromise their bearer's agility. Indeed, recent experimental evidence has demonstrated that, during routine or courtship swimming, males with longer swords expend more energy and consume more oxygen than males with shorter swords (Basolo and Alcaraz, 2003).

## Brood care in Jamaican land crabs

Jamaica is home to nine species of "land crab" in the family Grapsidae. These peculiar animals inhabit various terrestrial and freshwater environments on the island, and show varying degrees of dependence on water. Although many other species of grapsid crab exist elsewhere (primarily in intertidal communities around the world), what sets the Jamaican endemics apart, aside from their complete independence from the sea, is their exceptional parental devotion to larvae and juveniles. In most other crabs, hatched larvae are liberated into ocean waters to fend for themselves, but the Jamaican land crabs actively tend their broods.

In the Bromeliad Crab (*Metopaulias depressus*), for example, each mother raises her young in the axil of a water-filled bromeliad leaf. There, she circulates and thereby oxygenates the water, removes detritus, feeds her young, and protects them from predatory spiders and damselfly nymphs. She even hauls in empty snail shells, which provide a calcium source for her offspring and also serve as a pH buffer. In another Jamaican endemic, the Snail-shell Crab (*Sesarma jarvisi*), the adult either turns an empty snail shell upside down to collect rainwater, or actively carries water into the shell. For several ensuing months, the shell then becomes a nursery for the developing offspring.

Various morphological features are associated with these behavioral adaptations for brood care. For example, the Bromeliad Crab has a flattened body that allows it to squeeze into the narrow leaf axils of its host plants. The diversity of body plans and lifestyles among the endemic Jamaican crabs has raised questions about their

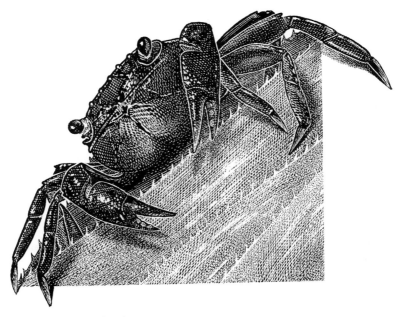

Jamaican Bromeliad Crab

evolutionary origins and phylogenetic affinities. One view, based primarily on morphological studies, held that some of the Jamaican land crabs might be related more
closely to grapsid species elsewhere in the world than to other land crab species
in Jamaica. In particular, one suspected group of evolutionary kin included freshwater species in the genus *Sesarmoides* from Southeast Asia. The competing view
held instead that all Jamaican land crabs evolved from the same ocean-inhabiting
ancestor (perhaps resembling *Sesarma curacaoense*, the only extant marine species
in Jamaica). In that case, the shared body plans of various Jamaican land crabs and
particular freshwater or terrestrial crabs elsewhere would be outcomes of convergent evolution.

To test these competing hypotheses, Schubart *et al.* (1998) assayed mtDNA
sequences from all living species of Jamaican land crab plus representatives of
other relevant crab species from Asia, and from Atlantic and Pacific waters flanking Panama. The resulting molecular phylogeny, shown in Fig. 4.8, implied the
following. First, all of the Jamaican land crabs belong to a single clade, meaning almost certainly that they are products of an adaptive evolutionary radiation
within or near the island. Second, traditional taxonomic assignments for grapsid crabs had not captured these evolutionary genetic relationships. For example,
Jamaica's Bromeliad Crab should have been placed in the genus *Sesarma* rather
than *Metopaulias*, by phylogenetic criteria. Third, the closest living relatives of

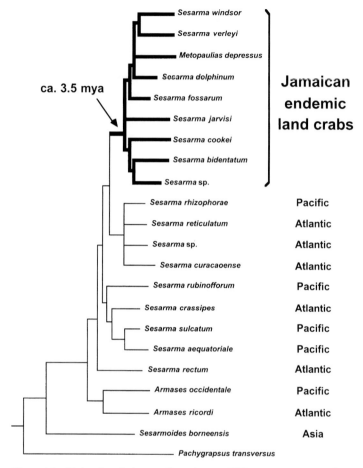

*Figure 4.8.* Molecular phylogeny (based on mtDNA gene sequences) for more than 20 species of grapsid crab (after Schubart *et al.*, 1998). Note the monophyly of Jamaican endemics, which appear to be descended from a shared ancestor dating to about 3.5 million years ago.

the Jamaican endemics appear to be marine intertidal species from the Americas (rather than Asia), meaning that the initial colonization of Jamaica's terrestrial realm was probably by a marine ancestor that inhabited the Caribbean.

The molecular data also permitted estimates of when that colonization event took place. First, a provisional molecular clock for mtDNA was calibrated for the crab genus *Sesarma* by comparing living species from opposite sides of Panama. This land barrier to marine organisms rose above the sea about three million years ago, and thereby produced sister pairs of species that ever since have been evolving independently in the tropical Atlantic versus Pacific Oceans. Based on a molecular clock calibrated from these sister species, the genetic distances observed among the Jamaican land crabs suggest that their evolutionary radiation began approximately 3.5 million years ago. These genetic inferences are compatible with geological evidence indicating that the Jamaican landmass became available for colonization only after the end of a mid-Tertiary inundation of that island by the Caribbean Sea about 10–20 mya.

Overall, results indicate that Jamaican land crabs evolved their diverse adaptations for non-marine life and complex brood care over the relatively short evolutionary span of just a few million years. By contrast, pairs of marine crabs in the Atlantic and Pacific Oceans, isolated by the Isthmus of Panama for roughly that same amount of time, have remained ecologically and morphologically very similar to one another. Thus, newly available terrestrial habitats in Jamaica, once successfully colonized by a marine crab ancestor, must have opened novel ecological opportunities that provided an evolutionary impetus for the rapid diversification of the island's descendent land crabs.

## Social parasitism of butterflies on ants

The life cycles of "large blue butterflies" (several species in the genus *Maculinea*) are nothing short of incredible. After hatching from tiny eggs, the early instars (larval stages between molts) feed for two or three weeks on the flower buds of particular host plants, such as members of the rose family. When the larvae reach the fourth instar stage, however, they drop to the ground and are picked up by particular ant species, normally red ants in the genus *Myrmica*. The ants carry the butterfly caterpillars into their nests, where, depending on the butterfly species, the caterpillars then adopt either of two distinct feeding habits: a predatory behavior in which they actively feed on grubs in the ant's brood; or a "cuckoo" behavior in which the butterfly caterpillars are fed regurgitated food, ants' eggs, or other prey items delivered mouth-to-mouth by worker ants in the colony. The nurse ants that do this feeding may get so carried away with the foster-parenting task that they neglect their own brood, even dicing up the colony's grubs and recycling them to feed to their caterpillar guests. After exploiting their ant hosts via either the predatory or the cuckoo-feeding routes, the caterpillars eventually go on to complete the life cycle by transforming into the familiar blue-winged adult butterflies that can be seen flitting about habitats in much of Europe and Asia.

How could such a bizarre relationship arise and persist between these butterfly larvae and their manipulated ant hosts? In terms of proximate mechanisms, some pieces of the puzzle are known. Most notably, each butterfly species has evolved distinctive hydrocarbon molecules that closely mimic those synthesized

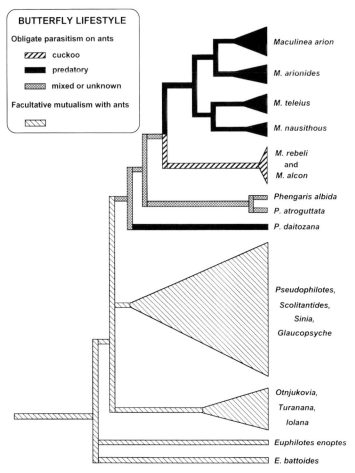

*Figure 4.9.* Molecular phylogeny for *Maculinea* butterflies and their allies, also showing the PCM-inferred evolutionary histories of their close behavioral interactions with ants (after Als *et al.*, 2004).

by its particular host species of *Myrmica* ant. When secreted by the caterpillars, these chemicals apparently act as super-stimuli that trick worker ants into taking the larvae into their nests and adopting them as if they were their own. In other words, butterfly larvae gain social acceptance into the ant colonies by pheromonal skullduggery, and then selfishly exploit their duped hosts. The relationship is sometimes referred to as an example of "social parasitism" because social settings are involved and the ants receive no apparent benefits from the association.

All of which still begs the question: by what sequence(s) of evolutionary events did such a remarkable set of cross-species relationships emerge? To begin to address this issue by using the logic of PCM, Als *et al.* (2004) estimated a molecular

Red ant tending the larva of a *Maculinea* butterfly

phylogeny (from nuclear and mitochondrial DNA sequences) for several species of *Maculinea* butterfly and their relatives with known or suspected host–guest lifestyles. The results of the analysis, summarized in Fig. 4.9, led the authors to several interesting conclusions. First, all of the *Maculinea* species belong to a mono-phyletic group that is phylogenetically embedded within a broader assemblage that also includes members of *Phengaris* (another recognized genus containing species that are obligate social parasites on ants). Second, the inferred distribution of butterfly lifestyles on this phylogeny implies that predatory social parasitism (rather than cuckoo-feeding social parasitism) was most likely the ancestral condition for the *Maculinea–Phengaris* clade, and that social parasitism itself probably evolved from earlier butterfly lineages displaying facultative mutualisms (opportunistic and jointly favorable interactions) with ants. Thus, obligate cuckoo feeding probably evolved from obligate social predation (rather than vice versa), and all forms of social parasitism in these butterflies and ants most likely emerged from earlier mutualistic associations.

A third finding from the PCM analysis was that several of the named species of predatory social parasites in *Maculinea* have relatively deep, internal phylogenetic subdivisions, whereas the two recognized *Maculinea* species (*rebeli* and *alcon*) with cuckoo-feeding lifestyles do not. Thus, cryptic species may well exist among the predatory but probably not the cuckoo-feeding species of social parasites. Further research, especially on details of chemical matches between ant hosts and their butterfly parasites, will be needed to further clarify species boundaries in these butterflies, but the current phylogenetic findings already have conservation implications. *Maculinea* butterflies are highly endangered throughout the Palearctic region and have become flagship invertebrate species for conservation efforts in Europe. If, as now seem likely, some traditionally described morphospecies (morphologically distinguishable forms) actually consist of two or more distinctive biological species (reproductively isolated populations) with different host specificities, then the true number of endangered species will inevitably be larger than previously supposed, and the total population size of each will be even smaller.

Although obligate social parasitism of *Maculinea* butterflies on ants may be an extreme evolutionary outcome, other kinds of cross-species relations with ants are widespread in nature. It is estimated that about 100 000 insect species have evolved one or another mechanism to coexist in close association with ants, including such adaptations as armor to resist attack, mimicry to avoid detection, or bodily secretions to behaviorally appease or feed ants (Hölldobler and Wilson, 1990). In many of these interactions, both partners benefit, as for example when aphids secrete honeydew (sugar-rich liquids) for their ant tenders in return for protection from predators. With such behavioral interactions being commonplace, it is not difficult to imagine that natural selection might also encourage cheating by one or both participants, and that any cheating mechanisms might often involve the same attributes (such as chemical cues) that otherwise permit mutualistic cohabitation. Such has apparently been the case for *Maculinea* butterflies, which selfishly exploit the largesse of their ant hosts.

## Of monkeyflowers and hummingbirds

Co-evolution is the joint evolution of two of more ecologically interacting species. Because tight functional connections often exist between co-evolving species, relevant phenotypic traits in such taxa are likely to have co-evolved as well. This means that co-evolving biological systems provide fertile ground for comparative PCM, as has already been illustrated in earlier essays dealing with predator–prey systems (see *Müllerian mimicry butterflies* in Chapter 3) and host–parasite interactions (see *Social parasitism of butterflies on ants*, above). In general, by mapping the phylogenetic origins and historical interconversions of multiple phenotypes onto independently generated molecular trees, new insights can emerge about which of those phenotypic traits may have causally influenced the co-evolutionary patterns among the interactants.

Another illustration of this comparative PCM approach comes from monkeyflowers (genus *Mimulus*, section *Erythranthe*) and their animal pollinators. Several species of monkeyflower occur in various regions of western North America, where they collectively display a great diversity of distinctive and often rather dichotomous floral traits. For example, depending on the species, nectar volumes may be low (0–1.1 µl per flower) or high (7–17 µl), petals may be erect or not, stamen and pistil lengths may be short (0–21 cm) or long (32–50 cm), and flower colors may be red, yellow, purple, pink, or white. Such beauty and variety have no doubt contributed to the popularity of monkeyflowers as model systems for numerous ecological studies.

Also of interest to ecologists are the co-evolutionary relationships of monkeyflowers and their pollinators. Several species are pollinated by hummingbirds, whereas others are serviced primarily by insects. Thus, several questions arise: is pollination by hummingbirds mono- or polyphyletic in the *Erythranthe*? Did the evolutionary origin(s) of hummingbird pollination phylogenetically coincide with the appearance of particular suites of floral traits, and if so, which ones? And what might such evolutionary correlations imply about the selection pressures involved in switches between alternative pollination systems?

To address these questions, Beardsley *et al.* (2003) first generated a molecular phylogeny for species of *Erythranthe* by using DNA markers, and then reconstructed on this tree (based on the logic of parsimony and maximum-likelihood PCM) the probable evolutionary transitions and co-transitions between alternative floral characteristics and pollinators. The results of their analyses are summarized in Fig. 4.10, from which the following conclusions emerged. First, pollination by hummingbirds arose on at least two separate evolutionary occasions in the genus *Erythranthe*: once in the common ancestor of a clade composed of *M. verbenaceus*, *M. eastwoodiae*, *M. nelsonii*, and *M. rupestris*, and again, independently, in a near ancestor of *M. cardinalis*. Second, the inferred evolutionary origins of several floral traits (red flowers, long stamens and pistils, and backward upper petals) coincided perfectly on the phylogenetic tree with the two evolutionary origins of hummingbird pollination. Third, other floral traits, such as nectar volume (Fig. 4.10) and presence versus absence of carotenoid pigments in the petals (not shown), did not co-map perfectly with hummingbird pollination on the *Mimulus* phylogeny.

These PCM-based conclusions also gave some fresh perspectives on the co-evolutionary games played by monkeyflowers and their pollinators. Some of the phylogenetic associations were expected. For example, it is well known that hummingbirds generally prefer red flowers, so the historical connection between red flowers and avian pollination came as no surprise. Nor did the phylogenetic

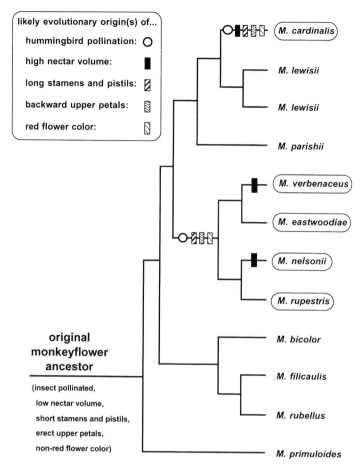

*Figure 4.10.* PCM analyses of several floral traits against a molecular phylogeny for about a dozen species of *Mimulus* monkeyflower, some of which are hummingbird-pollinated whereas the others are insect-pollinated (after Beardsley *et al.*, 2003). Note the phylogenetic coincidences in the tree between the inferred evolutionary origins of hummingbird pollination and the evolutionary origins of red flowers, long stamens and pistils, and erect petals.

correlation between hummingbird pollination and long flower parts: for plants serviced by small birds with long bills, extended pistils and stamens are important for the effective removal of pollen from anthers and the deposition of pollen on stigmas. Other phylogenetic patterns were unexpected, however. Notably, the less-than-perfect phylogenetic association between nectar volume and humming-bird pollination seemed puzzling. Hummingbirds usually prefer flowers that give them high nectar rewards (as food), so plants that attract hummingbirds usually have evolved flowers that are super-loaded with nectar. But two hummingbird-

pollinated monkeyflower species (*M. eastwoodiae* and *M. rupestris*) have among
the lowest nectar volumes of any *Mimulus* species.

One hypothesis for this peculiar situation is as follows. Perhaps *M. eastwoodiae*
and *M. rupestris* are taking advantage of hummingbirds by receiving pollination
services without bearing the added metabolic costs of high nectar production. In
theory, this selfish (rather than mutualistic) tactic by the plants should only be
practicable over a longer term if these species were rare relative to "honest" red-
flowered species that do offer hummingbirds high nectar rewards. Consistent with
this possibility, both *M. eastwoodiae* and *M. rupestris* are uncommon and geographically localized (indeed, *M. rupestris* is known from only one small population in
southern Mexico).

Overall, the PCM analyses by Beardsley and colleagues indicate that each of the
two bird-pollinated monkeyflower clades (Fig. 4.10) originally evolved from ancestral plants that were probably insect-pollinated and had short stamens and pistils,
erect upper petals, and relatively low nectar volumes. These PCM analyses further
showed that the evolutionary transitions from insect to hummingbird pollination
were historically associated with key alterations in all of these floral conditions
(with one partial exception being the low nectar volumes that either were evolutionarily retained or re-evolved in the lineages leading to M. *eastwoodiae* and
*M. rupestris*). The take-home message, again, is that comparative phylogenetic
approaches can complement direct contemporary ecological studies by illuminating co-evolutionary processes of the past.


## Parthenogenetic lizards, geckos, and snakes

Parthenogenesis is reproduction via an unfertilized egg, without genetic involve-
ment by males or sperm. The word itself comes from the Greek roots *genesis*
meaning production, and *parthenos* meaning virgin. Under parthenogenesis, an
unreduced egg (one whose chromosomal constitution is identical to that of the
mother) develops directly into an offspring who is thus genetically identical to
its one-and-only parent as well as to any siblings. Taxa that employ this clonal
reproductive mode usually consist solely of females. Despite its seemingly peculiar nature, parthenogenesis (or similar reproductive modes) is displayed by miscel-
laneous fishes and amphibians, as well as by scattered representatives of reptilian
groups ranging from lizards in the families Lacertidae, Xantusidae, and Agamidae to
particular geckos (Gekkonidae), chameleons (Chamaeleonidae), and blind snakes
(Typhlopidae). In whiptail lizards alone (Teiidae), more than a dozen parthenogenetic species are known.

Actually, the word "species" does not apply comfortably to parthenogenetic taxa because these organisms do not engage in sexual reproduction or experience normal genetic recombination. Thus, each parthenogenetic taxon is instead often referred to as a unisexual "biotype." Much is known about how these biotypes arise. In every known instance in vertebrates, parthenogenesis had its evolutionary origin in a hybridization event between related sexual species. For example, the unisexual biotype *Cnemidophorus uniparens* arose via interspecific hybridization between two sexual species of North American lizards: *C. burti* (the father in the original cross), and *C. inornatus* (the mother). The evidence is as follows: each surveyed specimen of *C. uniparens* houses a full collection of chromosomes (and, hence, of nuclear genes) from both *C. burti* and *C. inornatus*, yet it carries a mitochondrial genotype inherited maternally from *C. inornatus* only.

Parthenogenetic biotypes are of special phylogenetic interest for several reasons, not least of which concerns enigmas about evolutionary persistence. In sexual species, the normal profusion of recombinational genetic variation that arises each generation is crucial to evolutionary survival, affording such organisms the collective genetic scope to adapt to ever-changing environments. But within any parthenogenetic lineage, such genetic scope is negligible because all individuals are genetically identical (barring rare *de novo* mutations that may accumulate between successive generations of mothers and their clonal daughters). Thus, conventional wisdom holds that an absence of genetic recombination confers a short evolutionary lifespan on any unisexual biotype. Parthenogenetic biotypes clearly arise quite often (as judged by their wide taxonomic distribution), but seldom survive for long. If this speculation is correct, then parthenogenetic reproduction in extant taxa should characterize only some short twigs in the Tree of Life's outermost canopy. In other words, extant unisexual clades will seldom have persisted long enough to form significant phylogenetic branches, much less larger tree limbs or trunks.

Many empirical tests of this phylogenetic prediction have been conducted, typically based on the following molecular approach. First, the hybrid evolutionary origin of each unisexual biotype is deduced by using a combination of nuclear and mitochondrial genetic markers, as described above. Each monophyletic unisexual lineage thereby identified is then mapped onto the broader phylogeny of the sexual species to which it is allied. Phylogenetic relationships among the sexual species may be estimated from nuclear or mitochondrial genes, but phylogenetic placement of the unisexual taxon within this historical framework is normally based on mtDNA gene sequences. This is because mtDNA molecules, by virtue of their maternal inheritance, provide a common genetic yardstick for phylogenetic appraisals of all-female parthenogens as well as their sexual ancestors. Indeed, for

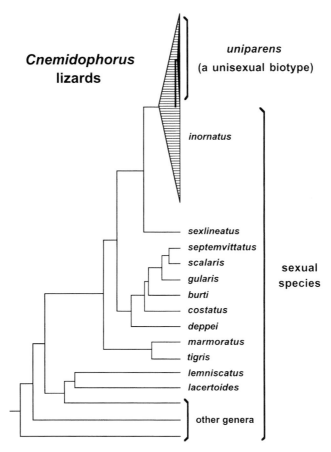

*Figure 4.11.* Molecular phylogeny estimated for *Cnemidophorus* lizards (after Dessauer and Cole, 1989). This depiction shows *C. uniparens* as deeply embedded within the *C. inornatus* clade.

parthenogenetic biotypes, which consist solely of females, the matrilineal phylogeny is in principle one and the same as the organismal genealogy (i.e. the single hereditary pathway traversed by all genes).

Figure 4.11 depicts one such phylogenetic analysis involving a parthenogenetic whiptail lizard (*Cnemidophorus uniparens*) and 12 related sexual species in the same genus. In this case, the broader phylogeny was estimated from nuclear genetic data, and the refined placement of the unisexual biotype was based on mtDNA. Clearly, *C. uniparens* arose very recently in evolution, as judged by the following observations: it occupies only one tiny branch within the broader matrilineal clade of *C. inornatus* (the sexual species that was its maternal progenitor); and *C. inornatus* itself is deeply embedded within the phylogeny of *Cnemidophorus* lizards.

Thus, *C. uniparens* is far younger than the genus *Cnemidophorus*, and indeed its origin considerably postdates whatever times of evolutionary separation (a few tens of thousands of years at most) distinguish maternal lineages of *C. inornatus* from other sexual species of *Cnemidophorus* lizard.

These kinds of phylogenetic patterns are typical of what has been discovered in comparable molecular examinations of more than 20 unisexual biotypes in various vertebrate groups. To pick a second reptilian example, a common parthenogenetic gecko (*Heteronotia binoei*) in Australia was found by molecular analysis to represent only a small phylogenetic subset of the matrilineal diversity of its sexual ancestor, implying that this unisexual biotype arose recently in evolution (probably at a single geographic site in the western part of the continent). Among all extant unisexual vertebrates similarly examined to date, one of the oldest well-documented biotypes appears to be the Mexican fish *Poeciliopsis monacha-occidentalis*, whose lineage was estimated from molecular evidence to be about 60 000 years old. But from an evolutionary perspective, even this amount of time is merely a brief evening gone.

Some parthenogenetic biotypes are common and widespread today. The gecko *H. binoei*, for example, is broadly distributed in Australia, and some of the unisexual *Cnemidophorus* lizards are common in deserts of the American southwest. Thus, at least some parthenogenetic vertebrates can experience considerable ecological success despite their extinction-prone clonal nature. However, molecular phylogenetic analyses have confirmed that, in nearly every case, any ecological good fortune that a unisexual biotype might enjoy is evolutionarily fleeting.

### Delayed implantation

A perennial challenge in evolutionary biology is to weigh the relative impacts of natural selection and phylogenetic inertia (or constraint) in determining the distributions of particular traits among contemporary species. One such reproductive character in mammals is delayed implantation (DI), a phenomenon in which an assemblage of postzygotic cells in a pregnant female developmentally arrests for an extended period of time before implanting into the uterine wall, where embryonic development then resumes. DI is a special case of embryonic diapause (ED), broadly defined as any mechanism by which a temporary cessation of embryonic development is achieved. ED is a widespread but poorly understood life-history strategy in various mammals, birds, fishes, insects, and plants. The great diversity of ED modes in the biological world suggests that this state of suspended animation has arisen multiple times and may have strong selective advantages in some ecological circumstances. The patchy distribution of ED across animal taxa also

Badger

suggests that either phylogenetic constraints or ecological conditions might have played key roles in determining where the trait now occurs.

In mammals, constitutive DI has been reported in more than 50 species representing seven taxonomic orders and ten families. It is especially common in carnivorous mammals in the families Ursidae (bears), Phocidae and Otariidae (seals), and Mustelidae (otters, weasels, skunks, badgers, and their allies). This latter family is of special interest because DI is well developed in some species but absent in others, and also because several species that display DI (e.g. the North American River Otter (*Lutra canadensis*), Stoat (*Mustela erminea*), and Western Spotted Skunk (*Spilogale gracilis*)) are supposedly close relatives of species that do not (European River Otter (*L. lutra*), Weasel (*M. nivalis*), and Eastern Spotted Skunk (*S. putorius*), respectively).

To further investigate the origins and evolutionary transitions between DI presence and absence in Mustelidae, Thom *et al.* (2004) conducted PCM using as backdrop an evolutionary tree that had been constructed by Bininda-Emonds *et al.* (1999). The results are summarized in Fig. 4.12, from which Thom and colleagues drew the following conclusions: DI is the probable ancestral (plesiomorphic) condition for mustelids, and multiple state changes between DI and no-DI are required to explain the current distributions of these traits (see also Lindenfors *et al.*, 2003).

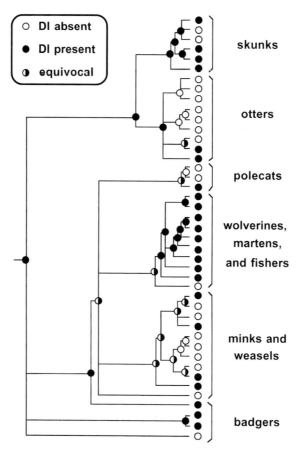

*Figure 4.12.*  Maximum parsimony reconstruction of ancestral character states for delayed implantation (DI) in 43 living species in the family Mustelidae (after Thom *et al.*, 2004).

Upon further evaluations, Thom *et al.* (2004) also discovered that DI is significantly more prevalent in mustelid species that inhabit higher latitudes, and in species that are longer-lived.

These latter two findings are of special interest because they relate to plausible adaptive scenarios for the evolution of DI. The basic idea is that DI may enhance individual fitness under any ecological circumstances in which natural selection favors organisms that are able to decouple the timings of mating and parturition. This may help to explain the observation that DI tends to be more prevalent in high-latitude species, because long winters and increased seasonality in these realms (as compared to more stable climatic regimes nearer the equator) may make it adaptively advantageous for females to mate at one time period (e.g. the autumn) but delay birth to a distant season (e.g. the spring) that may be much better suited for progeny survival. By similar reasoning, the association of DI with longevity may be hypothesized to arise because only in long-lived species do females have the luxury of delaying implantation. In general, in any ecological setting where the optimal period for finding high-quality mates and for giving birth differ, DI can be adaptively highly advantageous.

Lindenfors *et al.* (2003) offered a slightly different adaptive explanation for the distribution of DI. Noting that DI is probably the ancestral state for mustelids, and that it appears to be lost more commonly in species with small body size, they suggested that the evolutionary loss of DI relates to the relative costs of fecundity rather than timing of the mating and birth seasons *per se*. They argue in particular that DI has been selected against in small species where a prolonged gestation period is likely to impose relatively higher fecundity costs. Factors such as unpredictable mortality, multiple litters per year, and briefer lifespans are all known to be positively correlated with smaller body size, and thus are likely to impose relatively higher fecundity costs on smaller compared with larger species.

All of these adaptive explanations tend to invoke natural selection as a primary agent driving mustelid diapause. They do not, however, eliminate the possibility that phylogenetic inertia is influential too, as is perhaps suggested by the perpetuation of DI (and of no-DI) across successive nodes in several portions of the mustelid phylogenetic tree (Fig. 4.12). From these and additional considerations, Thom *et al.* (2004) were led to conclude that, although ecological factors can predict the distribution of DI in extant mustelids, phylogenetic constraint is likely to have played an important role as well.

ED is merely one example of a life-history feature that has proved to be relatively labile in evolution, having shifted back and forth over rather short phylogenetic timeframes. Other such examples discussed in this book include host specializations in brood-parasitic birds (see *Egg dumping and foster parentage*), avian nesting behaviors (*The avian nest*), aspects of live bearing in reptiles (*Egg laying and live bearing*) and in pregnant fishes (*Piscine placentas*), and transitions between larval life-forms in marine invertebrates (Chapter 5, *Dichotomous life histories of marine larvae*). This kind of evolutionary lability implies that natural selection has often been a powerful force in shaping many life-history adaptations.

# 5 | More behaviors and ecologies

The behaviors of organisms evolve just as surely as do their physical features. Indeed, the connections between a species' behavioral repertoire and its suite of morphological attributes are almost always so tight as to blur distinctions between organismal form and function, i.e. between what a creature is and what it does. Adaptive co-evolution between a species' behavioral ecology and its physical attributes accounts for why we don't observe, for example, vegetarian leopards or predatory antelopes.

As presaged in Chapter 4, PCM analyses can be conducted on behavioral and lifestyle traits just as they can on morphological ones. This chapter will provide several additional examples ranging from evolutionary analyses of the kangaroo's bipedal hop to the organization of multi-species lizard communities on Caribbean islands, and from how pufferfish gained the ability to inflate their bodies into anti-predation balls to how particular bacteria sense the earth's geomagnetic field. In most of the following case studies, PCM analyses have been applied as well to anatomical attributes associated with particular organismal behaviors, meaning that the topics explored in this chapter will also overlap to some degree with those covered in Chapter 2.

## The kangaroo's bipedal hop

When in fast gait, kangaroos and their relatives (family Macropodidae) employ the bipedal hop. Indeed, two-legged hopping is mandatory for rapid cross-country locomotion because these animals' forelimbs are short and weak whereas their long hindlimbs are powerfully constructed for propulsion. The evolutionary emergence of bipedal locomotion was accompanied by several other changes in macropodid anatomy, such as a reduction in the numbers of hind toes (from five to three or four, giving a more hoof-like structure), and a thickening of the tail for balance and stability. The evolutionary transition from quadrupedality to bipedality started more than 50 million years ago when a possum-like marsupial (family

Gray Kangaroo and Musky Rat-kangaroo

Phalangeridae) with arboreal tendencies gradually took up an increasingly ter-restrial lifestyle. Today, the characteristic bipedal hop is how nearly all species of Macropodidae bound around the Australian continent. Traditionally, macropodids are divided into two taxonomic subfamilies: Macropodinae, with approximately 40 living species of kangaroos, wallabies, and pademelons; and Potoroinae, with about 10 species of rat-kangaroos, potoroos, and bettongs.

Included somewhere in this evolutionary menagerie is the enigmatic Musky Rat-kangaroo (*Hypsiprymnodon moschatus*). This animal looks superficially like a small macropodid, but it likes to climb on fallen trees and branches, and when running it bounds on four legs rather than two. In several other anatomical features, this quadruped is effectively intermediate between the presumed possum-like ancestor of macropodids and the modern bipedal hoppers. So where does it fit in the marsu-pial evolutionary tree? Based on some anatomical details, taxonomists traditionally had considered the Musky Rat-kangaroo to be a sister taxon to the Potoroinae, as shown in panel A of Fig. 5.1. If so, this would imply either that bipedality evolved twice within the Macropodidae (once each in the respective ancestors of Macro-podinae and Potoroinae), or that bipedality evolved only once at the base of the

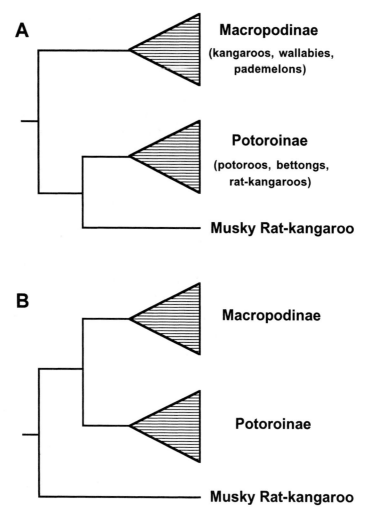

*Figure 5.1.* Two alternative hypotheses for the phylogenetic position of the Musky Rat-kangaroo (after Burk *et al.*, 1998). Scenario B appears to be favored by recent molecular evidence.

macropodid tree and that quadrupedality was then regained in the *H. moschatus* lineage.

Recent molecular data have painted a different phylogenetic picture, however. According to phylogenetic analyses of mtDNA nucleotide sequences, the Musky Rat-kangaroo is instead a sister taxon to Macropodidae, having initially separated from the proto-macropodid lineage approximately 45 million years ago, or about 15 million years before the ancestral separation of Macropodinae and Potoroinae. This revised phylogenetic arrangement (panel B of Fig. 5.1) suggests that bipedal

hopping arose only once in macropodid evolution and that there have been no evolutionary reversions to quadrupedality (at least among the marsupial lineages that
still survive today). The phylogenetic distinctiveness of *H. moschatus* has also led
to calls that this species be placed in its own taxonomic family (Hypsiprymnodon-
tidae).

   This new phylogenetic scenario has further ramifications. It helps to make sense,
for example, of the observation that several postcranial features of *H. moschatus*
are intermediate between the possum-like ancestor of macropodids and those of
the derived bipedal hoppers. The evolutionary polarities of particular anatomical
features have become clearer as well. For example, the Musky Rat-kangaroo has
a simple stomach like that of the phalangerid marsupials in Australia, but which
is quite unlike the complex forestomach (with special adaptations for digesting
cellulose) of kangaroos and other macropodid species. This implies that the simple stomach is the ancestral condition, such that the derived complex stomach
provides another defining phylogenetic feature of the Macropodidae. Another
derived feature uniting the true Macropodidae may be single-offspring litters.
Unlike the twinning habit of *H. moschatus* (and the multi-offspring litters of various other marsupials), all macropodids typically produce just one offspring at a
time.

   Scientists can only speculate about the specific sequence of environmental challenges that eventuated in the singular evolution of bipedalism from quadrupedal-
ism in Australian macropodids some 50 million years ago. Perhaps bipedal hopping became increasingly adaptive as the early proto-macropodid moved from the
rainforest floor to the open savannahs that became far more prominent during the
initial aridification of the Australian continent. From this perspective, *H. moschatus* would simply have retained a suite of ancestral characteristics that have kept it
well suited for its preferred rainforest habitat.

   Another interesting question is why bipedal hopping became the preferred
mode of cursorial locomotion in Australian mammals, whereas elsewhere in the
world quadrupedal galloping evolved in many placental mammals such at horses
and antelopes. At least two possibilities have been suggested. Perhaps certain
aspects of foot anatomy of the possum-like ancestor predisposed or even constrained macropodid evolution to a bipedal rather than quadrupedal trajectory
(Marshall, 1974). Or perhaps the evolutionary potential for the marsupial fore-
limb was restricted by the role of neonatal crawling (Szalay, 1994). Unlike a developing placental baby, safely housed and nurtured within its mother's womb, a
marsupial newborn must use tiny dexterous limbs and paws to grab and grope
its way from birth canal to the marsupial pouch where it will complete its early
development.

Wahlberg's Epauleted Fruit Bat

## Powered flight in winged mammals

Molecular phylogenies are often most interesting when they can serve to referee and adjudicate apparent disputes between seemingly conflicting lines of morphological or other evidence on historical relationships. One such phylogenetic quarrel began in the mid 1980s when John Pettigrew published a report in the prestigious journal *Science* documenting previously unsuspected electrophysiological and neurological similarities between fruit bats (suborder Megachiroptera, or "mega-bats") and members of the Order Primates (which includes lemurs, lorises, tarsiers, monkeys, and apes). In particular, neuroanatomical features connecting the eye's retina to the midbrain were remarkably similar in representatives of these two taxonomic groups, yet different in key details from the presumed ancestral pattern displayed by most other mammals including the Microchiroptera (typical nocturnal bats, or "micro-bats"). Pettigrew interpreted these observations as strong evidence that mega-bats and primates have a tighter phylogenetic connection than do mega-bats and micro-bats (as summarized in panel A of Fig. 5.2).

This provocative suggestion, which became known as the flying-primate hypothesis for mega-bats, flew directly in the face of conventional wisdom, which held that mega-bats and micro-bats (traditional Chiroptera) are one another's closest living relatives, with primates being a more distant phylogenetic outlier

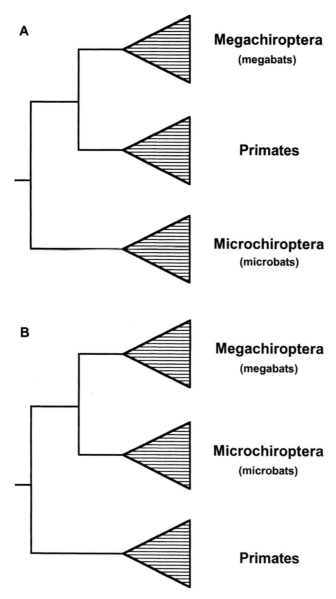

*Figure 5.2.* Two alternative hypotheses regarding phylogenetic relationships of bats and primates. Recent molecular evidence generally supports some version of scenario B.

(panel B of Fig. 5.2). A flurry of evaluative activity followed, including phylogenetic re-examinations based on DNA sequences from a variety of mitochondrial and nuclear genes. The reason that the issue attracted so much attention is that most systematists formerly supposed that powered flight arose only once in mammalian evolution (in the presumed common ancestor of a monophyletic Chiroptera). If

Pettigrew's hypothesis was correct, however, powered flight arose once in the micro-bat line and again, independently, in a mega-bat subset of the ancestral primate lineage. For this reason, the flying-primate scenario is also referred to as the diphyletic hypothesis for wing-powered mammalian flight.

From other morphological and behavioral evidence, the diphyletic notion is not easy to dismiss. Apart from powered flight and its associated apparatus (such as highly modified forelimbs, and membranous flaps of skin that stretch from the bat's hands and arms to its body and hind legs), the Megachiroptera and Microchiroptera seem superficially quite different. For example, many micro-bats (of which there are nearly 1000 living species) can lower their body temperature and may hibernate for long periods, whereas mega-bats lack this physiological ability. Furthermore, micro-bats are mostly small, nocturnal insect-eaters that use ultrasonic echolocation to help avoid obstacles and catch prey, whereas the mega-bats are large, diurnal, visually oriented fruit-eaters.

Thus, according to the diphyletic hypothesis, the capacity for flapping flight in micro-bats and mega-bats is a stunning example of convergent evolution rather than an indication of shared ancestry. Clearly, the argument goes, flight can be highly adaptive, so it is reasonable to suppose that micro-bats and mega-bats evolved this remarkable behavioral capability independently. By contrast, under the monophyletic hypothesis for Chiroptera, flight is a valid synapomorphy for micro-bats and mega-bats, so any stunning evolutionary convergence must be in the neurovisual pathways shared by mega-bats and primates. Good sight can be highly adaptive, so under this view mega-bats and primates probably experienced selective pressures that independently forged similar neuroanatomical pathways for exquisite vision.

Only secure independent evidence (not directly tied to flight capacity or visual acuity) is likely to reveal the true historical relationships among micro-bats, mega-bats, and primates, and hence whether convergent evolution occurred with respect to flight or visual acuity. Such phylogenetic evidence has emerged from molecular data. For example, Bailey *et al.* (1992) compared the sequences of a globin nuclear gene and found that, whereas 39 derived nucleotide substitutions were uniquely shared by the mega-bat and micro-bat species assayed, only three such changes were shared by mega-bats and primates. Support for the monophyletic status of Chiroptera has also emerged from sequence comparisons of several other nuclear and mitochondrial genes. Indeed, some molecular studies indicate that micro-bats are paraphyletic with respect to mega-bats, meaning that the latter are merely a phylogenetic subset of the broader micro-bat + mega-bat clade (and perhaps also implying that echolocation was secondarily lost in the Megachiroptera). Thus, in the light of current molecular evidence, the flying-primate hypothesis has been rejected in favor of the monophyletic origin for powered mammalian flight.

## Magnetotaxis in bacteria

Some bacteria have a built-in compass that helps them orient to the earth's geomagnetic field as they swim. The working parts of this compass are magnetosomes – iron-rich particles encased by membranes – that create a magnetic dipole within the bacterial cell. As a sensory device, an internal compass affords a bacterium a greater ability to find and maintain its physical position with respect to chemical or other environmental gradients that may be important to its survival.

The magnetosomes in most magnetotactic bacteria consist of an iron oxide compound ($Fe_3O_4$), but iron sulfide compounds ($Fe_3S_4$ or $FeS_2$) form the magnetosome particles of a few recently discovered bacterial taxa inhabiting sulfidic brackish-water environments. Although both the iron-oxide and iron-sulfide types of bacteria sometimes can be found in the same general areas, their chemical micro-habitats (in terms of oxygen and sulfide concentrations, for example) appear to differ. Quite likely, the magnetosomes play a key role in helping the bacteria navigate and properly situate themselves in suitable environments.

Magnetotaxis is a remarkable adaptation, but did it arise only once or multiple times during bacterial evolution? A recent analysis by DeLong and colleagues (1993) provisionally answered this question by examining the phylogenetic arrangements of iron-oxide and iron-sulfide magnetosomes. The backdrop for this PCM exercise was an estimate of bacterial phylogeny based on nucleotide sequences of a slowly evolving gene encoding small-subunit ribosomal RNA (a key component of the cellular apparatus that translates nucleic acids to proteins). Based on their analysis (Fig. 5.3), the authors concluded that magnetotaxis in bacteria probably had at least two ancient and independent evolutionary origins, one in the so-called α-subdivision and the other within the δ-subdivision of Proteobacteria. Furthermore, these separate geneses seem to correspond precisely to the dichotomy between iron-oxide and iron-sulfide magnetosomes. These phylogenetic findings make considerable sense because, most likely, the biochemical bases for biomineralization and magnetosome formation are fundamentally different in the iron-oxide and iron-sulfide types of bacteria.

The broader lessons are twofold. First, PCM analyses can be extended even to some of the planet's smallest inhabitants. Second, even such "simple" creatures as bacteria display remarkable genetic ingenuity, in this case having convergently evolved behaviorally similar but biochemically distinct solutions to the sophisticated task of geomagnetic orientation.

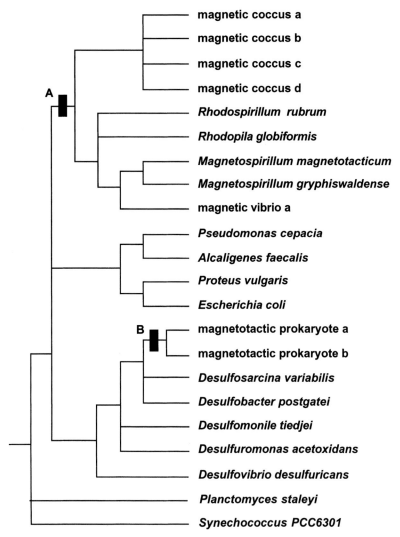

*Figure 5.3.* Molecular phylogeny (maximum parsimony analysis of ribosomal RNA gene sequences) for representative bacteria, provisionally documenting independent evolutionary origins for iron-oxide (A) and iron-sulfide (B) magnetotaxis (modified from DeLong *et al.*, 1993).

## Cetacean origins

Especially when creatures make an evolutionary transition to a very different ecological niche or lifestyle, the remodeling of their anatomies, physiologies, and behaviors can be so severe as to obscure their phylogenetic origins and affinities. Such is the case for whales, porpoises, and their allies (Cetacea). Biologists have long understood that cetaceans colonized the oceans from terrestrial ancestors, perhaps about 50 million years ago, but exactly what those ancestral creatures were, and which surviving lineages are related most closely to extant cetaceans, have been among evolution's most enduring mysteries.

The challenge in understanding whales' phylogenetic affinities stems in large degree from the wholesale reorganization of body plans attendant with the transition to a fully aquatic existence. For example, extant cetaceans are unique among living mammals in lacking hindlimbs (with the exception of an internal pair of rod-like vestiges). Other evolutionary adaptations to open-water marine life include a fusiform body superficially like that of fish, a shortened neck but telescoped skull, a reduced pelvic girdle, additional vertebrae, and nostrils that open on top of the head (where they form blowholes for breathing air). In effect, natural selection for an aquatic lifestyle has acted like a giant historical eraser, effacing much of the morphological, physiological, and behavioral evidence that might otherwise have helped to illuminate cetacean phylogeny.

None the less, by examining anatomical details and fossils, systematists began more than a century ago to pare down lists of candidate kin. The near-unanimous conclusion was that cetacean's closest relatives are to be found within the ungulates (hoofed mammals). This still left tremendous latitude for phylogenetic speculation, however, because ungulates are a highly diverse assemblage, often subdivided into several distinctive taxonomic orders. Among extant forms, these include the Artiodactyla (even-toed or cloven-hoofed species ranging from pigs and hippopotamuses to cattle, deer, and camels), Perissodactyla (odd-toed forms such as horses, tapirs, and rhinos that bear their weight on a middle toe), and Proboscidea (elephants), among others. Additional phylogenetic analyses based on molecular data eventually refined the list of contestants further, convincingly showing that whales' closest living kin lie somewhere within the Artiodactyla.

This still left considerable room for phylogenetic speculation, however, because artiodactyls themselves are a highly diverse lot traditionally subdivided into several large groups (taxonomic suborders and families). Among extant forms, the three major groups are as follows: Ruminantia (ruminant animals with a modified stomach housing bacteria that digest plant material) including 34 species of deer (Cervidae), about 140 species of goats, sheep, cattle, gazelles and allies (Bovidae), Pronghorn Antelope (Antilocapridae), giraffe and okapi (Giraffidae), chevrotains or mouse-deer (Tragulidae), and others; Tylopoda including camels and llamas ("pseudo-ruminant" animals that practice a physiologically different form of rumination); and the non-ruminant Suiformes including pigs and peccaries. Most of the recent systematic research on these groups has involved sorting out molecular phylogenetic relationships of various taxa, with particular attention devoted to identifying the creatures to which cetaceans are related most intimately.

Therein lay a great surprise: hippopotamuses (traditionally placed in Suiformes) may be cetacean's nearest living kin. Support for this contention initially came from DNA sequence data from several nuclear and mitochondrial genes, but an even more compelling line of evidence was the discovery that several diagnostic genetic elements (specific types of retroposon that during evolution jump into particular sites in the nuclear genome but almost never excise back out) are uniquely shared by whales and hippopotamuses, presumably by virtue of their inheritance from a common whale–hippo ancestor who first acquired those elements. The taxonomic distributions of additional retroposons of this sort further fleshed out the phylogenetic picture for other artiodactyl groups as well (see Fig. 5.4). Thus, molecular data not only confirmed that artiodactyls and whales are historically linked, but they also raised the distinct possibility that cetaceans are nested deeply inside the artiodactyl clade (which therefore became known as the Cetartiodactyla clade). This finding in turn has prompted serious reconsideration of at least two conventional notions: that mesonychians – long-extinct hoofed carnivores that superficially resembled massive warthogs – were the immediate ancestors of whales; and that hippopotamuses are phylogenetically embedded within the Suiformes.

The discovery that extant cetaceans and hippopotamuses may be sister taxa is intriguing for another reason as well. Hippos and whales share several special aquatic adaptations including lack of hair, an absence of sebaceous glands, and the use of underwater vocalizations in communication. Traditionally, these features were presumed to have evolved independently in hippos and whales, from unrelated ancestors. However, if the phylogenetic picture currently emerging from molecular evidence is correct, then these behavioral and morphological adaptations might be synapomorphies instead, and therefore genuinely reflective of common ancestry. This exciting but unorthodox hypothesis will warrant additional investigation from both molecular and morphological points of view.

### Feeding and echolocation in whales

Systematists traditionally subdivided the Cetacea (whales, porpoises, and allies) into two supposedly distinct monophyletic groups: the Odontoceti (echolocating toothed forms) with about 67 species, and the Mysticeti (filter-feeding baleen whales) with 10 species. Because extant sperm whales (*Physeter macrocephalus* and *Kogia breviceps*) have teeth and echolocate, most experts thought they were closely allied to (Fig. 5.5A) or perhaps even embedded within the odontocete clade. It came as quite a surprise, therefore, when initial molecular evidence from mitochondrial as well as some nuclear genes indicated that sperm whales are phylogenetically closer to baleen whales than to other toothed cetaceans (Fig. 5.5B). In other words,

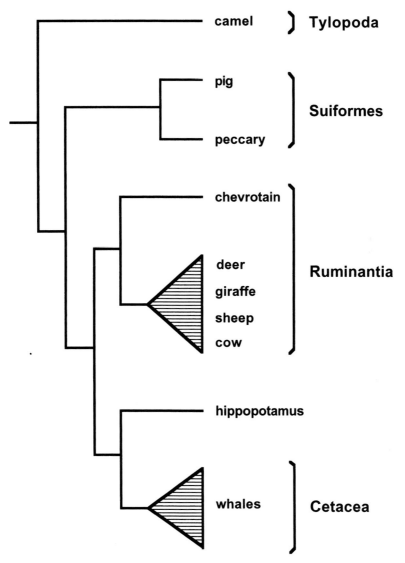

*Figure 5.4.* Phylogeny estimated from recent molecular evidence for cetaceans and particular artiodactyls (modified from Nikaido *et al.,* 1999).

the Odontoceti (as traditionally defined) appeared to be a paraphyletic rather than a monophyletic group. This little phylogenetic adjustment, seemingly trivial at first glance, has prompted wholesale reinterpretations of the evolutionary histories of several key behavioral and morphological features in the Cetacea.

   With regard to feeding adaptations, the new phylogenetic arrangement suggested that teeth-presence was probably the ancestral cetacean condition (making it a

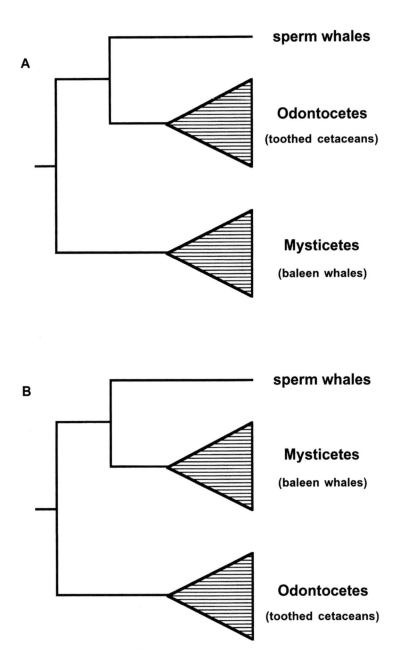

*Figure 5.5.* Two alternative hypotheses regarding the phylogenetic relationship of sperm whales to other cetaceans (see text).

**external blowholes, leading to nasal passages**

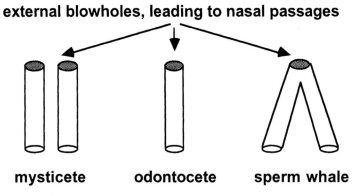

**mysticete        odontocete        sperm whale**

Blowhole forms in whales

symplesiomorphy, and hence uninformative for concluding that toothed cetaceans are monophyletic), and that baleen feeding was a shared-derived trait (a synapomorphy) correctly identifying a mysticete clade. This hypothesis also appeared plausible from other evidence. Namely, although adult mysticete whales lack teeth and instead use baleen (comb-like plates growing from the palate) to filter vast volumes of ocean water for plankton, their embryos do display rudimentary teeth early in development, suggesting still incomplete loss of an ancestral condition. Furthermore, various extinct cetaceans in the Suborder Archaeoceti, from which all modern whales and porpoises are probably descended, also were fully toothed. These phylogeny-based inferences bolstered a notion that the evolution of filter feeding in a mysticete ancestor was a key innovation that permitted cetaceans to exploit a novel and abundant food source. This in turn permitted the evolution of such impressively large animals as the Blue Whale (*Balaenoptera musculus*)(the heaviest animal, reaching 150 tonnes, ever to have inhabited the planet).

Blowholes (external nares, or nasal openings) are another anatomical feature reinterpreted under the new phylogenetic framework. All extant baleen whales have two blowholes, whereas all toothed whales (including sperm whales) seem to have just one. In this case, the obvious ancestral condition for Cetacea is two external nares, as in other mammalian groups (including the terrestrial ungulates from which cetaceans arose). At face value, the derived one-nare state might seem to be strong evidence that sperm whales belong to the odontocete clade, but closer inspection revealed that the sperm whale's blowhole leads immediately to two internal nasal passages (see drawing above), such that the full respiratory layout probably resembles that of mysticetes more closely than that of odontocytes. This all seemed to make more sense in light of the new molecule-based phylogeny.

Finally, active echolocation (the use of sonar) is believed to be characteristic of all toothed cetaceans (including sperm whales), but supposedly never evolved in the baleen whales. This too seemed at face value to contradict the molecular phylogeny, but again a further evaluation suggested otherwise. According to Milinkovitch (1995), echolocation may be the ancestral condition for whales and dolphins, meaning that its shared presence in sperm whales and other toothed cetaceans reflects a symplesiomorphy not necessarily indicative of a monophyletic status for these species. This hypothesis seemed plausible for at least two reasons. First, an acoustical melon (a fatty lens in the forehead that is an important component of the echolocation system of toothed whales) has been described in vestigial form in baleen whales, suggesting that it was present more fully in the common ancestor of the Mysticeti. Second, from general evolutionary principles, it is normally much easier to lose a complex adaptation than to gain one, so the loss of complex echolocation capabilities and associated anatomical structures in baleen whales may not be too surprising (especially given their plankton-feeding rather than large-prey-capturing lifestyles).

None the less, most of the conclusions reached above remained provisional pending further molecular and morphological analyses. Such re-examinations were not long in coming. Notably, Nikaido *et al.* (2001) soon published an analysis of the taxonomic distribution of a series of small genetic sequences known as SINEs (short interspersed elements) that are believed to provide powerful phylogenetic markers because particular ones arise rarely in genomes but once present are almost never lost. Remarkably, these evolutionary "SINE-posts" supported the more traditional notion that toothed whales (including sperm whales) are probably monophyletic after all. According to the SINE evidence, the sperm whale lineage was an early offshoot just barely on the Odontoceti side (rather than the Mysticeti side) of the cetacean family tree (as in Fig. 5.5A). If true, then most of the morphological and behavioral reinterpretations prompted by the earlier molecular findings will have to be revisited yet again.

Such is the nature of scientific give-and-take often encountered in phylogenetically difficult biological settings. A broader point none the less emerges from the various molecular studies on cetacean relationships. Collectively, they provide a wonderful example of how even tiny alterations in the apparent structure of a phylogenetic tree can profoundly affect evolutionary interpretations of numerous behavioral and morphological adaptations mapped onto that tree. This sensitivity is a double-edged sword: PCM-based conclusions can be revelational, but they can also be highly sensitive to any errors in the phylogenetic reconstructions themselves.

## The phylogeny of thrush migration

Every autumn, an estimated 10 billion birds of nearly 400 species disappear from northern climes only to reappear in smaller numbers the following spring. In centuries past, naturalists (such as Aristotle) were uncertain whether these birds had traveled elsewhere, or hibernated. We now know that they migrate, often on epic journeys requiring navigational and exertional feats that nearly defy human comprehension. In the New World, many of these Neotropical migrants travel to Central or South America before returning to North America to breed the following spring, whereas the wintering destination for most European and Asian migrants is the African continent.

Migration is costly (energetically and in terms of travel risks), so the compensatory benefits must also be substantial. Ornithologists traditionally viewed the advantages of migration to birds in either of two complementary ways: as an avoidance of severe environmental challenges (such as a harsh climate and shortage of insects) during high-latitude winters; or as a positive exploitation of predictable but temporary ecological opportunities (such as long days and an abundance of food) during high-latitude summers. These two classes of explanation can yield different interpretations about the evolutionary origin(s) of migration to and from northern climes. Under the "escape" scenario, migrants are interpreted as ancestral northerners who started to migrate during times of climatic deterioration, such as for example during Ice Ages of the Pleistocene Epoch; whereas under the "exploitation" scenario, migrants are viewed as ancestrally tropical birds that evolved migrational tendencies to capitalize upon abundant summer resources farther north. Depending on the taxonomic group under consideration, both of these scenarios probably have elements of truth.

Far more certain is that many avian taxa can gain or lose migratory behaviors very rapidly, as evidenced by different tendencies often displayed by species that are phylogenetically closely related (see below). On occasion, major changes in migratory habit have even been documented by direct observations. For example, Mediterranean populations of the European Serin (*Serinus serinus*) are mostly sedentary, but populations newly established in northern Europe in the past century have already become migratory. Conversely, a now-resident population of Fieldfares (*Turdus pilaris*) in Greenland was settled recently by migratory birds from Europe; and, some Barn Swallows (*Hirundo rustica*) now nest in Argentina rather than return to the Northern Hemisphere like most of their conspecific brethren. So, migratory behaviors of birds often show remarkable evolutionary plasticity. On the other hand, historical legacies are evident as well. For example, a new breeding

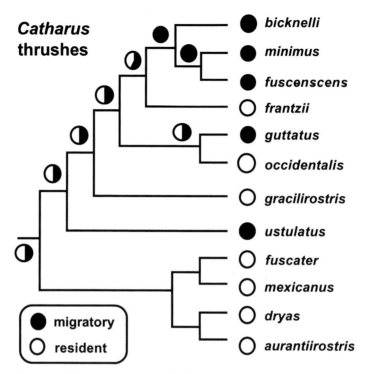

*Figure 5.6.* Phylogeny estimated for *Catharus* thrushes from mtDNA sequence data (after Outlaw *et al.*, 2003). Pie diagrams show the relative probabilities of two alternative character states (migratory or resident) at each node.

population of Pectoral Sandpipers (*Caladris melanotos*) recently became established in Siberia, but rather than migrate south through the Orient as do most other Siberian sandpiper species (as would make better geographic sense), these birds instead return to South America (via Alaska), as did their immediate ancestors. Similarly, some Northern Wheatears (*Oenanthe oenanthe*) colonized Greenland from the British Isles, but like their ancestors they still return (via Europe) to wintering sites in Africa, rather than migrating more directly south into the Americas.

Such apparent tensions between evolutionary plasticity and historical legacy in avian migration indicate that evolutionary dissections of these behaviors must be conducted taxon-by-taxon, and that PCM approaches should help. These propositions were recently put to test on *Catharus* thrushes. This New World genus (in Muscicapidae) consists of 12 closely related species, five of which are long-distance migrants between North America and Latin America, and seven of which are year-round residents in the tropics. Figure 5.6 shows a molecular phylogeny for these dozen species, and also summarizes a most-parsimonious reconstruction of migratory behaviors onto that mtDNA-based tree.

Several points emerged from this analysis. First, migratory behavior in *Catharus* thrushes appears to be polyphyletic (as indicated by the fact that artificially forcing migratory species into a clade resulted in a statistically worse phylogeny). Second, when the phylogeny was interpreted in conjunction with species' ranges and other evidence, it seemed likely (but not conclusive) that the original thrush ancestor was a southern-resident species, and that migratory behavior to northern latitudes evolved several times. Finally, it was also true, however, that the statistical probabilities of migratory versus resident behaviors at many internal nodes in the tree (see the pie diagrams) seldom permitted definitive conclusions about the evolutionary trajectories of migratory behaviors. An alternative possibility, for example, is that migration was the ancestral condition for thrushes, and that the habit was then lost on different occasions. The broader point is that when any traits are evolutionarily highly labile (as avian migratory habits are thought to be), then deducing their precise phylogenetic histories over substantial evolutionary time may simply not be very tractable.

## Pufferfish inflation

About 150 living species of pufferfish (Tetraodontidae) and spiny puffer (Diodontidae) get their names from a remarkable defensive behavior in which they literally inflate their bodies into balloons. When threatened by a predator (or hooked by a fisherman), a pufferfish gulps mouthfuls of water that are pumped into an expandable stomach. Quickly, the fish assumes a grossly bloated physique that makes the animal difficult for a predator to attack or swallow. In the spiny puffers, bony spines in the skin stand erect when the fish is inflated, further enhancing this defense. After the threat has passed, a puffer releases water and reassumes its normal, somewhat more svelte condition.

This peculiar inflation behavior was made possible by several evolutionary modifications of the puffer's body. Unlike most fishes, modern puffers have highly stretchable skin on their lateral and ventral bodies, an extremely distensible stomach lining, no ribs (these would be an impediment to shape changes during inflation), and several structural peculiarities of the head, buccal cavity (mouth) and pectoral girdle (shoulder joint) that play key roles in water pumping. Biologists have long been intrigued by how non-puffer fish evolved into pufferfish, i.e. by what may have been the ancestral condition as well as subsequent intermediate stages in the evolution of full-blown inflation as a defensive behavioral tactic.

Wainwright and Turingan (1997) addressed these issues by plotting relevant anatomies and behaviors onto a phylogenetic tree for pufferfish and their relatives. Puffers belong to the order Tetraodontiformes, which also includes

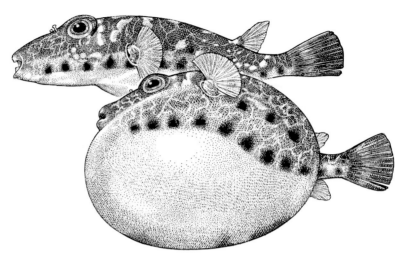

Southern Pufferfish, normal and inflated

spikefishes (Triacanthodidae), triplespines (Triacanthidae), triggerfishes (Balisti-
dae), and filefishes (Monacanthidae), among others. General body outlines of fish
representing these groups are shown in Fig. 5.7. The Tetraodontiformes themselves
are one of approximately 40 taxonomic orders of bony fishes. Both deep and shallow
portions of the piscine phylogeny proved to be relevant to the authors' following
postulate of a four-stage series of evolutionary transformations for pufferfish inflation (see Fig. 5.7).

   The hypothesized sequence of evolutionary steps begins with a buccal compression behavior common to virtually all fish species: generalized coughing. This
forceful action expels unwanted matter from the mouth, and is often used during feeding to eject indigestible parts of prey (such as the hard exoskeleton of a
sea urchin, or the tough skin of a worm). From this ancestral condition then came
tetraodontiform coughing, which is a more forceful and focused expulsion of water
related to the small mouth aperture and reduced opercular (gill-cover) openings
characteristic of all Tetraodontiformes. A third evolutionary step, anterior water
blowing from the buccal cavity, follows directly from the second. Most but not all
species of Tetraodontiformes (see Fig. 5.7) blow strong jets of water out of their
mouths to expose buried prey (e.g. in sand), manipulate prey items, clean prey
fouled by sediments, or in a few cases to assist in nest construction. Finally came
the inflation behavior itself, in which water pumped from the buccal cavity travels
in a posterior rather than anterior direction. Puffers achieve this outcome simply by
closing their mouths during buccal compression, thereby redirecting water through
the esophagus and into the stomach. Wainwright and Turingan also analyzed (in

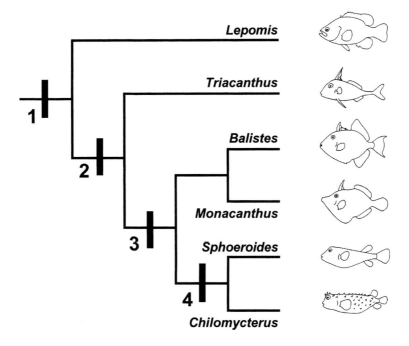

1. **generalized coughing**
     large mouth, laterally expandable buccal cavity, immobile
          pectoral girdle

2. **tetraodontiform coughing**
     small mouth aperture, reduced opercular opening, etc.

3. **water blowing**

4. **water inflation**
     novel mechanism of buccal expansion and compression,
          stretchy skin and stomach, mobile pectoral girdle, etc.

*Figure 5.7.* A partial phylogeny for tetraodontiform fishes including representative puffers (*Sphoeroides*) and spiny puffers (*Chilomycterus*), and also showing sunfish (*Lepomis*) as an outgroup (after Wainwright and Turingan, 1997). Unlike nearly all other phylogenies presented in this book, this tree was estimated from morphological data (a defensible practice in this case because many of the approximately 100 anatomical traits analyzed were presumably unrelated to the pufferfish inflation phenomenon *per se*, and hence afforded an independent estimate of organismal phylogeny). Plotted onto the phylogeny are four successive steps in a likely scenario for the evolution of pufferfish body inflation (see text).

great anatomical detail) numerous muscular and other evolutionary changes that apparently had accompanied this full sequence of events leading to pronounced body inflation.

The tetraodontiform phylogeny facilitated, indeed enabled, this evolutionary reconstruction indicating how various anatomies and associated behaviors in these fishes constitute a nested set of increasingly specialized adaptations. Generalized coughing, specialized coughing, water blowing, and water-based body inflation all share a common functional basis, yet through a succession of intelligible steps they have evolved to assume biological roles that now differ substantially among extant tetraodontiform species.

### Eusociality in shrimp

Human beings are not the most organized, social, or selfless creatures on the planet. That honor may belong instead to eusocial hymenopteran insects (ants, bees, and wasps). Eusociality is a complex suite of behaviors with the following characteristics: coordination and cooperation among individuals in the care of young; reproductive division of labor in colonies where sterile workers serve reproductive individuals; and overlapping generations of colony workers. In the eusocial hymenopterans, for example, female workers dutifully build and maintain the nest and rear the offspring of one or more queens. Eusociality is of special interest to biologists not only because it entails exceptionally high levels of societal organization, but also because it involves reproductive self-sacrifice by colony workers.

Eusociality is quite rare in the biological world. Apart from the social hymenopterans, it is known to occur in termites, a few species of gall-forming thrips, some clonal aphids, some social beetles, and burrow-dwelling naked mole-rats in the genus *Heterocephalus*. From an ecological perspective, species predisposed to evolve eusociality typically fall into one of two categories (Queller and Strassmann, 1998): "fortress defenders," which live inside a nest or protected site (a valuable resource that is both possible and necessary to defend as a group); and "life insurers," which may forage in the open but none the less benefit from group behaviors because overlapping generations of adults are needed to successfully care for young within the nest. The scattered distribution of eusociality among diverse animal taxa is *prima facie* evidence that the phenomenon is broadly polyphyletic.

From a genetic point of view, many aspects in the evolution of eusociality may best be explained by Hamilton's (1964) inclusive fitness theory. The basic idea is that colony members can make the evolutionary transition to extreme altruism (reproductive self-sacrifice, in this case) only when they are closely related genetically. Although an individual worker may not reproduce personally, its genes (including those that cause altruistic behavior) can become well represented in the next generation by virtue of kin selection in a smoothly operating colony of close relatives.

In addition to ecological and genetic perspectives on evolutionary transitions to eusociality, a complementary class of explanations can emerge from historical (i.e. phylogenetic) analyses. Marine shrimps nicely illustrate this point. An exciting and unanticipated recent discovery was that some sponge-inhabiting shrimp species in the genus *Synalpheus* display advanced eusociality. In these species, closely related individuals (often full sibs) live together, sometimes by the hundreds, inside a large sponge, and a lone female accomplishes most if not all of the colony's reproduction. By contrast, behavioral arrangements in other *Synalpheus* species range from asocial pair bonding to subsociality to various degrees of communality. This rich heterogeneity of social organization has afforded fine fodder for PCM reconstructions.

For example, a phylogeny (Fig. 5.8) for more than a dozen *Synalpheus* species showed quite unequivocally that advanced eusociality arose on at least three separate occasions in these animals (Duffy *et al.*, 2000). This discovery in turn enabled Duffy and colleagues to reconstruct the likely evolutionary histories of eusociality by making several independent phylogenetic contrasts between sister lineages displaying various gradations of social organization. These comparative analyses in conjunction with ecological data revealed that colonies of highly social species normally had many more individuals within a host sponge than did less social species, and that highly social taxa tended to share their host sponges with fewer congeneric species. These results were interpreted to support the idea that intense competition has acted as a primary selective pressure favoring the evolutionary elaboration of sociality in these (and perhaps other) animal lineages. In other words, during the evolutionary transitions to eusociality in sponge-inhabiting shrimps, progressively higher levels of cooperation among close genetic kin (and advantages of inclusive fitness for non-reproductive workers) probably enhanced colony success by making critical nesting sites easier to acquire and hold.

In principle, similar PCM analyses might be conducted on ants and termites, for example, but such efforts to date have been mostly frustrated by the ancient origins of eusociality in these groups, and the paucity of robust phylogenies for relevant lineages that might be intermediate in degrees of social organization (but see Danforth *et al.* (2003) for an interesting case study on primitively eusocial bees). By contrast, sponge-dwelling shrimps were highly favorable for PCM analyses because, within this monophyletic assemblage, eusociality clearly arose relatively recently and repeatedly in several evolutionary lines that collectively show rich variation in social make-up.

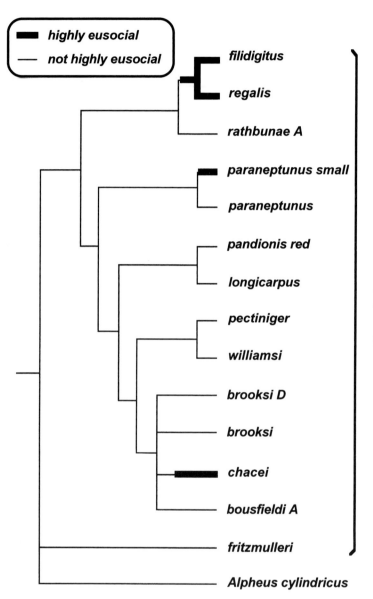

*Figure 5.8.* Phylogenetic tree, based on a combination of morphological and molecular data, for sponge-dwelling shrimps in the genus *Synalpheus* (after Duffy *et al.*, 2000).

## Evolutionary reversals of salamander lifecycles

Introductory textbooks sometimes oversimplify by portraying evolutionary lineages as always progressing from the less refined conditions of lowly ancestors to the more advanced phenotypes of higher descendants. For example, the initial colonization of land by fish-like proto-amphibians, and later the adaptive expansion into arid environments by amphibian-like proto-reptiles, are often portrayed as natural or inevitable progressions from relatively simple aquatic animals to the more sophisticated creatures of terrestrial realms. Such implications should be resisted, however, for several reasons: evolution has no inevitable trends (except, perhaps, that extinction is the ultimate fate of all lineages); descriptive terms such as "lowly" and "advanced" entail value judgments that may be difficult to defend by objective criteria; organisms in any environment can show exquisitely refined adaptations to their respective habitats; and all lineages alive today have been equally ingenious in the sense that each has found a way to survive evolution's trials and tribulations across the eons.

For these reasons, terms such as "advanced" or "higher" (if they are to be used at all) should be taken to imply merely that the organisms or traits in question evolved later (i.e. more recently) than others to which they are being compared. In this value-neutral sense, a widespread notion among herpetologists is that direct development is an advanced state of affairs compared with a biphasic lifecycle. Under direct development, eggs are deposited on land and embryos develop without the need for standing water (i.e. hatched embryos develop directly into juveniles and adults without passing through an aquatic larval stage). By contrast, under a biphasic lifestyle, eggs are laid in or near standing water and hatch into free-living aquatic larvae that later transform into terrestrial adults. The biphasic lifecycle is characteristic of many but not all amphibians. It can be viewed as functionally intermediate between the fully aquatic lifecycle of fishes and the fully terrestrial lifecycle of reptiles and other creatures with cleidoic eggs (see Chapter 4, *The chicken or the egg?*).

However, some amphibians display direct development. This terrestrial lifestyle appears to have evolved independently, from an ancestral biphasic lifecycle, in subsets of each of the three primary groups of living amphibians: anurans (frogs and toads), caecilians (legless worm-like forms), and urodeles (newts and salamanders). Among the urodeles, for example, many members of the species-rich Plethodontidae (lungless salamanders) lay their eggs on land and altogether forego an aquatic larval stage. Direct development was a key evolutionary innovation that helped free these skin-breathing salamanders from a reproductive dependency on standing water, thereby enabling them to better colonize and exploit terrestrial habitats.

The family Plethodontidae was conventionally subdivided into two taxonomic subfamilies: Plethodontinae, species of which exhibit direct development; and Desmognathinae, with a biphasic lifecycle. Typically, plethodontine species complete their entire life cycle in moist woodlands, whereas the larvae of nearly all desmognathine species must enter aquatic habitats to survive and grow. Traditionally, Plethodontinae and Desmognathinae were thought to be sister taxa, with the biphasic lifecycle presumed to be the ancestral plethodontid condition. Under this notion, biphasy was a primitive lifecycle from which direct development subsequently emerged as a more refined condition that enabled the plethodontine lineage to radiate adaptively in woodland habitats.

Recent molecular data have tweaked some aspects of this conventional wisdom. Based on phylogenetic analyses of combined mitochondrial and nuclear DNA sequences, Chippindale *et al.* (2004) reported that the desmognathine clade is embedded within (rather than sister to) the plethodontine assemblage (Fig. 5.9). In other words, plethodontines as traditionally defined are paraphyletic with respect to the desmognathines. As summarized in Figure 5.9, these findings in turn imply that free-living aquatic larvae and the biphasic lifecycle probably re-evolved in the desmognathines from the direct-development lifecycle of their immediate ancestors. This is an excellent example of how a trait can simultaneously be ancestral and derived, depending on the specified frames of reference in a phylogenetic hierarchy. In this case, direct development in the plethodontines is undoubtedly a derived condition in the broader context of amphibian phylogeny, but it is also an ancestral condition relative to the desmognathines who have secondarily lost this predecessor state. Conversely, aquatic larvae and the biphasic lifecycle are immediately derived characters in the desmognathines, but are ancestral conditions in a broader amphibian context. These shifting perspectives are attributable to a previously unsuspected evolutionary reversion from direct development to a biphasic lifecycle in a specific branch of the plethodontid phylogeny.

More than 20 desmognathine species inhabit the Appalachian streams of eastern North America, where they often overlap in range with similar numbers of plethodontine species. The reinvasion of aquatic habitats probably enabled the desmognathines to exploit an open niche or adaptive zone in a geographic region already packed with terrestrial salamanders. One evolutionary clue as to why the desmognathines were able to revert to aquatic lifestyles, whereas some other salamander groups did not (see the Bolitoglossini in Fig. 5.9), comes from embryological considerations. All members of the desmognathine clade (including their ancestors with direct development) happened to retain a hyobranchial apparatus (gill bar) that is a key feature of aquatic respiration by salamander larvae. This preadaptation for a watery lifestyle may have been a morphological precondition that enabled

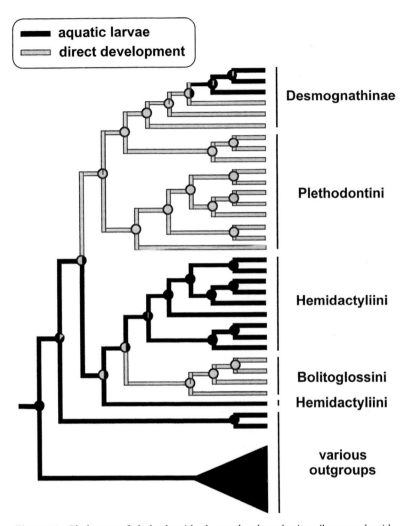

*Figure 5.9.* Phylogeny of plethodontid salamanders based primarily on nucleotide sequences from mitochondrial and nuclear DNA (after Chippindale *et al.*, 2004). Black branches indicate lineages deemed likely by PCM to have been biphasic in larval development (i.e. showing an aquatic phase); gray branches indicate lineages with direct development (i.e. no aquatic larval stage). Pie charts indicate the relative likelihoods of these two alternative lifestyles as reconstructed at various internal nodes in the tree.

desmognathine larvae to return to streams and thereby escape the presumably intense competition with direct-developing salamanders on land. Thus, the evolutionary retreat by desmognathines to a primitive biphasic lifecycle was also an extraordinary evolutionary advance that was key to their recent ecological success in Appalachian streams.

The biphasic lifecycle of desmognathine salamanders provides another example of how a seemingly complex and potentially adaptive phenotype can sometimes be lost and then regained during the evolutionary process, i.e. of how Dollo's law can sometimes be violated. Other such examples in this book can be found in *Snails' shell shapes* and *Winged walkingsticks* (in Chapter 2).

## Dichotomous life histories of marine larvae

Larval lifestyles in many taxonomic groups of marine invertebrates can be characterized as falling into one or the other of two distinct categories: lecithotrophy, in which larvae are non-feeding and receive nutrition from the yolks of relatively large eggs; and planktotrophy, in which larvae come from small eggs with no food stores, and instead feed actively while adrift in the sea. Suites of physical, behavioral, and developmental features are associated with these alternative life histories. For example, owing to their pre-provisioned food supply, lecithotrophic larvae typically have shorter developmental times and a simplified morphology (complex feeding adaptations are not required), and, by being rapidly growing and less mobile, have more limited dispersal capabilities than do planktotrophic larvae. These features in turn have evolutionary ramifications with regard to clutch sizes, magnitudes of species' geographic ranges, and rates of gene flow among conspecific populations (all of which are usually greater in planktotrophic species), and speciation rates (which often tend to be higher in clades with lecithotrophic larvae).

For most marine invertebrate groups, conventional wisdom has been that planktotrophy was the ancestral lifestyle from which lecithotrophy evolved recurrently. For example, PCM analyses showed that lecithotrophy probably evolved from planktotrophy at least four independent times in asterinid starfish (Hart *et al.*, 1997). Such tendencies can be rationalized by supposing that complex feeding structures, once lost by a lineage, are difficult to regain in evolution. On the other hand, probable instances have been identified (e.g. in littorinid and calyptraeid snails) in which larval feeding re-evolved in non-feeding lineages (Reid, 1990; Collin, 2004). Overall, a clear finding from phylogenetic analyses of several invertebrate phyla (including Mollusca, Echinodermata, and Annelida) is that modes of larval development can switch rapidly in geological time, and hence are not always tightly constrained evolutionarily.

Indeed, a number of instances are known in which lecithotrophy and planktotrophy co-occur as life-history alternatives *within* a taxonomic species. This situation, known as poecilogony, is of special scientific interest because evolutionary switches between larval modes in effect have been "caught in the act." Phylogenetic studies of one such situation, involving the polychaete worm *Streblospio benedicti*,

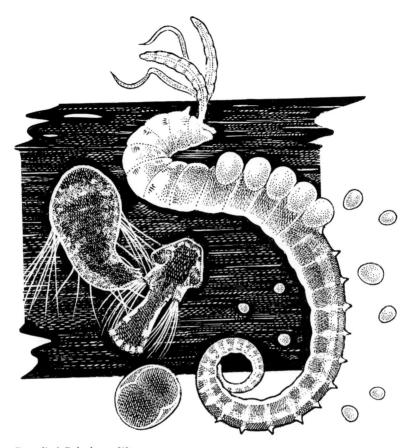

Benedict's Polychaete Worm

provide a nice example of how PCM analyses can be extended to the micro-evolutionary level as an aid in dissecting recent evolutionary shifts in life-history tactics.

Schulze *et al.* (2000) used mtDNA sequences to estimate an intraspecific phylogeny for several North American populations of what had traditionally been classified as *S. benedicti*, and then plotted the occurrences of three alternative larval developmental modes on the tree (Fig. 5.10). Most females in populations along the Atlantic Coast display pouch-based planktotrophy in which hundreds of small eggs are initially stored in dorsal brood pouches, but from which planktotrophic larvae emerge. Most females from the Gulf of Mexico also release large numbers of planktotrophic larvae, but in this case the small eggs had been stored in their gills (gill-based planktotrophy). By contrast, females on the Pacific Coast store just a few large eggs in a dorsal pouch, from which lecithotrophic larvae emerge

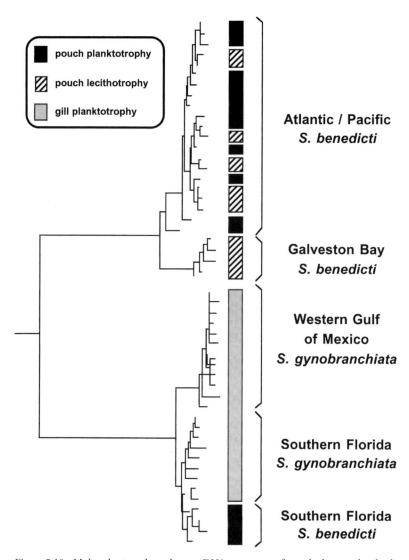

*Figure 5.10.* Molecular tree, based on mtDNA sequences, for polychaetes that had originally been described as conspecific (after Schulze *et al.*, 2000). Also shown adjacent to the tree are occurrences of alternative larval developmental modes and revised species designations (see text).

(pouch-based lecithotrophy). This latter life-history mode is also displayed by smaller numbers of females in the Gulf of Mexico and along the Atlantic Coast.

Several conclusions were drawn from these and additional empirical observations. First, as judged by the shallow evolutionary depths within each of the two major clades in the phylogenetic tree (Fig. 5.10), changes in larval life-history mode as well as brood-structure morphology can occur very rapidly, so these features must be evolutionarily highly plastic in these polychaetes. Second, experimental studies have indicated that switches between these larval life histories cannot be induced by altering environmental conditions such as food, temperature, or photoperiod, thus implying that these developmental modes are genetically hardwired. Third, the close phylogenetic interspersion of lecithotrophy and planktotrophy (and also of pouch-brooding versus gill-brooding) in portions of the tree made it impossible to determine with certainty whether evolutionary mode-switches in any one direction had been more common than those in another. Fourth, because one deep split was evident in the mtDNA phylogenetic tree, and because it distinguished most surveyed specimens in South Florida and the Gulf of Mexico from those collected elsewhere, suspicions were raised that a rather ancient (*c.* 10 million years ago) biogeographic separation had taken place (and also that recent range expansions or human-mediated transplantations might account for occasional departures from this basic phylogeographic pattern).

This fourth point also raised another issue, regarding species boundaries. In several other marine invertebrate species that had initially been characterized as poecilogenous, rigorous re-examinations revealed instead the presence of cryptic or sibling species, suggesting in turn that switches in larval developmental mode might themselves precipitate reproductive isolation and speciation. How do reproductive barriers map onto the phylogeny summarized in Fig. 5.10? Fortunately, *Streblospio* polychaetes can be reared and mated in the laboratory, and experimentally tested for reproductive compatibility. Such analyses by Schulze and colleagues have demonstrated that two reproductively incompatible units (and hence two biological species) exist within what formerly was considered *S. benedicti*. Furthermore, these two biological species (one of which has been renamed *S. gynobranchiata*) conform quite well, albeit not perfectly, to the two fundamental clades registered in the mtDNA phylogenetic tree (Fig. 5.10).

## Adaptive radiations in island lizards

A general goal in many PCM analyses is to weigh the relative roles of natural selection ("selective determinism") versus idiosyncratic vagaries of the past (historical contingencies) in having shaped the present-day arrangements of particular biological characteristics. As this section will attest, sometimes these analyses can be conducted even at the level of ecological communities.

*Anolis* lizards are a conspicuous element of the vertebrate fauna on various islands of the Caribbean. Often, several morphologically distinctive species co-exist on a given landmass but differ in habitat use. For example, anole

Cuban Brown Anole

assemblages on each of the four major islands of the Greater Antilles (Cuba, His-
paniola, Jamaica, and Puerto Rico) include species representing at least four of
the following "ecomorphs" (each named for the type of microhabitat normally
occupied): grass–bush, twig, trunk–ground, trunk, trunk–crown, and crown–giant.
The present-day co-existence of multiple lizard species on a given island undoubtedly is facilitated by these niche separations, but how the ecomorphs came to be
is another issue, one that can be informed by phylogenetic analyses.

One possibility is that each ecomorph type evolved only once and subsequently
spread to multiple islands either via past colonization events or by ancient vicariance (i.e. if a common ancestor of each ecomorph inhabited an Antillean landmass
that later was sundered into separate islands). If so, species representing a given
ecomorph class should be one another's closest living kin, regardless of which
islands they now occupy. Alternatively, each ecomorph type might have evolved
independently on various islands. Under this hypothesis, species inhabiting a given

island often might be one another's closest evolutionary relatives, regardless of which ecomorph categories they represent.

To decide between these competing possibilities, Jonathan Losos and his colleagues (1998) used mtDNA sequences to estimate a phylogeny for more than 55 species of Caribbean *Anolis* lizard that had been characterized for ecomorph membership. With few exceptions, members of the same ecomorph class from different islands proved not to be closely related, and no ecomorph category constituted a monophyletic lineage relative to the others. Instead, species representing various ecomorphs were more-or-less randomized among branch-tips of the phylogenetic tree. These findings showed conclusively that similar sets of *Anolis* ecomorphs evolved recurrently on the Caribbean islands. The PCM reconstructions further suggested that many (at least 17) lineage transitions among ecomorph classes had occurred during the evolutionary process, and that the historical sequences of these transitions probably differed from island to island (Fig. 5.11). For example, species representing the trunk–crown and twig ecomorphs proved to be sister taxa on Cuba, but each had closest relatives among the crown–giant or trunk–ground ecomorphs on the other islands surveyed.

The interspersion of ecomorph types throughout the molecular phylogeny of *Anolis* contrasts diametrically with how the species group with one another in terms of their general morphological appearances and microhabitat preferences. In these latter features (which tend to reflect niche dimensions), all species representing each ecomorph clustered together in a phenogram regardless of which Caribbean islands they now inhabit (Losos *et al.*, 1998).

Collectively, these findings indicate that natural selection has generally been of greater impact than historical contingency in shaping the present-day ecological communities of *Anolis* lizards on Caribbean islands. Almost irrespective of the island and whatever ancestral forms may have been present in the inoculatory lineages, natural selection (perhaps largely via competitive interactions) has molded the lizard communities into consistent sets of specialized ecomorphs that effectively subdivide the available resource base. Thus, although the vagaries of history are still registered in the molecular genetic makeup of *Anolis* lizards in the Caribbean, the determinism of natural selection has been a major player in the adaptive radiations and community compositions of these animals.

## Spiders' web-building behaviors

Like bird nests (see *The avian nest*, Chapter 4), spider webs are extended (outside-the-organism) phenotypes. Also like bird nests, spider webs display great architectural variety that has arisen from genetic changes across evolutionary time in the animals' innate design and construction behaviors. Thus, spider webs themselves,

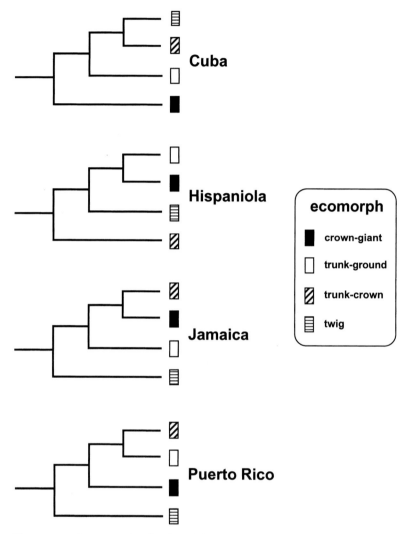

*Figure 5.11.*   Summary of molecular phylogenetic relationships among species of *Anolis* lizards representing four distinct ecomorph types that cohabit each of four Caribbean islands (after Losos *et al.*, 1998). Note that the phylogenetic topologies for these ecomorphs differ from island to island.

like avian nests, can properly be said to evolve, just as do endogenous phenotypes such as organismal anatomies or biochemistries.

   The silk of spiders, produced in abdominal glands, is a proteinaceous substance that is emitted as a liquid but quickly hardens into elastic strands of incredible strength (as evidenced, for example, by the fact that silk fibers of some tropical spiders are used by native peoples for catching fish). As a spider extrudes silk

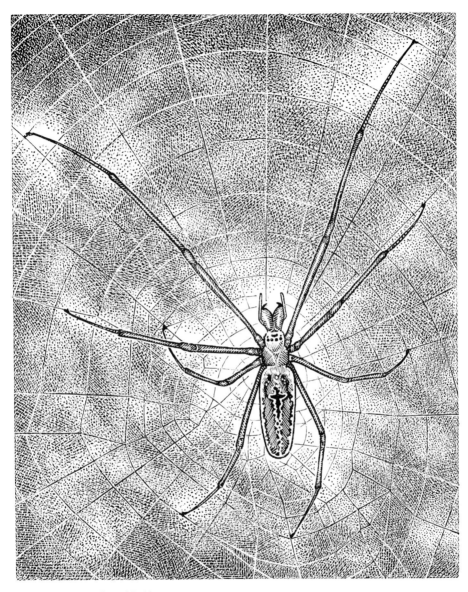

Long-jawed Spider

from several duct openings near its anus, six or eight tiny modified appendages known as spinnerets craft the threads into often elaborate webs. The architecture of each web is taxon-specific and stereotypical. Across species, webs range from haphazard draglincs and simple meshes, to sophisticated orbs and funnels, to grand ornate networks that surely must be listed among nature's most beautiful works of art.

However, spider webs are designed not for esthetics, but for function: primarily to catch insects and other invertebrate prey. This raises a fundamental question. Do the different web designs in any particular taxonomic group of spiders reflect mostly the varied functional demands of contemporary life in contrasting ecological settings, or, alternatively, do they track phylogeny mostly irrespective of the spiders' current ecological circumstances? In other words, are web architectures fluidly molded by natural selection, or are they severely constrained by historical legacies?

To begin to address these questions, Blackledge and Gillespie (2004) gathered mtDNA sequences from several Hawaiian orb-weaving spiders in the genus *Tetragnatha*. They then used the resulting estimate of molecular phylogeny (Fig. 5.12) as an evolutionary canvas for picturing and interpreting the relative impacts of natural selection and historical constraint on web design. Among the surveyed species, webs come in three basic architectures that are distinguished by size (large versus small) and/or by thread density (many supporting radial fibers and small mesh widths between spiral elements, versus fewer radial fibers and larger mesh widths). When plotted adjacent to the molecular phylogeny, each of these distinctive web motifs proved to be displayed by extant species situated in different *Tetragnatha* lineages and clades (Fig. 5.12). For example, large webs with dense threads were shared by Maui's *T. stelarobusta* and Hawaii's *T. hawaiensis*, despite the fact that these two species occupy very distant branches in the evolutionary tree.

From this PCM exercise, Blackledge and Gillespie concluded that each ethotype (i.e. each category of web-building behavior as evidenced by a distinctive web architecture) probably had two or more independent origins in the evolutionary history of Hawaiian *Tetragnatha*. Thus, convergent evolution in complex web-building behaviors was common, suggesting an important role for natural selection as a governing force in web design. Indeed, there was little evidence for phylogenetic constraints on interconversions among web designs, as further reflected in the fact that even closely related sister taxa (such as *T. stelarobusta* and *T. eurychasma*) sometimes build webs of different styles.

Additional research will be required to document the particular form of selection that quite evidently has played an important role in shaping the genetic underpinnings of web design. Several hypotheses will need to be tested by observation or experiment. For example, different species of *Tetragnatha* spider are believed to specialize on different subsets of insect prey, so one obvious hypothesis is that natural selection based on prey availability has led to predictable web designs in particular kinds of habitat. A second possibility involves natural selection on spiders via habitat-specific predation risks, which might vary predictably as a function of web architectures. In the Hawaiian archipelago, this hypothesis seems unlikely, however, because many common predators of mainland spiders are absent from

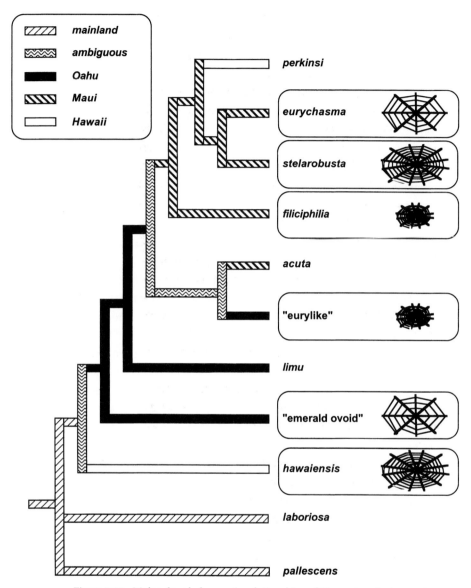

*Figure 5.12.* Molecular phylogenetic relationships of 11 species of orb-weaving spider in the genus *Tetragnatha* (after Blackledge and Gillespie, 2004). Also shown are species' ranges and web designs.

the islands, and because the most likely Hawaiian predators (honeycreeper birds) hunt during daylight hours when these nocturnal spiders have removed their webs. A third hypothesis is even more intriguing: that sexual selection may have played a role in shaping the evolution of web design. Courtship and mating in *Tetrag-natha* take place on the web itself, so perhaps mating preferences help to drive web architecture. If this is true, then one of the ecological functions of web design might be esthetics after all – not to human eyes of course, but rather to the many (usually eight) discriminating eyes of each potential spider mate.

### Lichen lifestyles

Symbiosis (from the Greek *syn* meaning together, and *bios* meaning livelihood) is a term used to describe pairs of species that live together without harming one another. It can include neutralism (0, 0), but usually is reserved for situations in which at least one party benefits: commensalism (+, 0), obligatory mutualism (+, +), or non-obligatory protocooperation (+, +). A common sentiment is that mutually beneficial forms of cooperation often tend to emerge gradually from neutral or antagonistic associations during the co-evolution of ecologically interacting species.

One of life's most remarkable symbioses involves the intimate association between an alga and a fungus to form a lichen. In this mutualistic joining of forces, the fungus normally provides the supportive tissue whereas the alga supplies food via photosynthesis. However, the alga of some lichens also can live without a fungus, and vice versa, so some lichens may reflect protocooperation rather than strict mutualism. Furthermore, in the fungal–algal alliance that is a lichen, the possibility of negative interactions between the participants is not excluded, such as if, for example, the fungus parasitically digests some of the algal symbionts on occasion. Indeed, among free-living fungi that might be phylogenetically allied to various lichen-forming taxa, both algal-pathogenic species and saprobic species (those that gain nutrients from dead or decaying matter) are well known.

Collectively, there are many types of lichen, meaning that different fungi and algae have entered into the associations. To examine the fungal part of the evolutionary equations, Andrea Gargas and colleagues (1995) plotted the lichen lifestyle onto a molecular phylogeny (based on ribosomal DNA sequences) for several dozen species representing two major fungal taxonomic groups, the Ascomycetes (yeasts, molds, morels, and truffles) and the Basidiomycetes (smuts, rusts, and mushrooms). The goals were twofold: to identify the minimum number of independent fungal origins for the lichen lifestyle, and to address whether the mutualistic symbioses had evolved from more parasitic types of interactions.

The results of this PCM analysis are summarized in Figure 5.13. They show quite conclusively that fungi have entered into symbiotic associations with algae on at least five separate evolutionary occasions: three times in the basidiomycete clade and twice in the ascomycete clade. Regarding the former, two of the three assayed "basidiolichens" (*Multiclavula mucida* and *Omphalina umbellifera*) produce

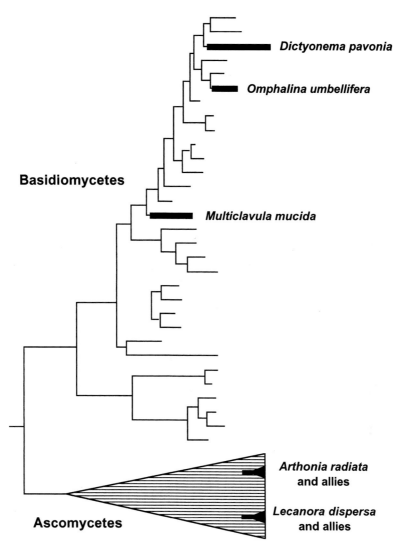

*Figure 5.13.* Five independent evolutionary origins of lichen lifestyles (heavy black lines and clades) as deduced from a PCM analysis based on a molecular phylogeny for about 70 fungal species (after Gargas *et al.*, 1995). Included in the PCM analyses were extant lineages representing more than 30 surveyed genera in Basidiomycetes (lineages shown in detail above) and more than 40 genera of Ascomycetes (lineages shown in summary format in the clade below). In this depiction, *Arthonia radiata* and *Lecanora dispersa* (and their allies) are shown as deeply embedded within the ascomycete clade.

fruiting structures that are alga-free and much like those of their non-lichen-forming mushroom relatives, suggesting in these cases that relatively few evolutionary changes may have attended the transitions to symbiosis. The third fungal basidiolichen is represented by *Dictyonema pavonia*, a close relative of wood-rotting fungi. It has haustoria (parasitic threads or hyphae) that invaginate the cells of its algal associates. With regard to the "ascolichens" (see Fig. 5.13), *Lecanora dispersa* also forms intimate haustorical connections to its algal associate, and among its close evolutionary allies are many saprobic fungal forms. Finally, the lichen-forming *Arthonia radiata* has hyphae that even more aggressively penetrate the symbiotic algal cells, so this species acts much like a parasite as well as a host. Furthermore, *Arthonia* is evolutionarily allied to fungal groups that include virulent plant pathogens.

The study by Gargas *et al.* (1995) underscored a long-suspected notion that the category "lichen," although ecologically meaningful, is not a coherent unit in terms of phylogenetic origins. It also made clearer the idea that the lichen lifestyle is a derived condition that emerged from multiple types of ancestral association, and that the progression is not always from aggressive parasitism to friendly mutualism. Although saprobic and pathogenic fungi sometimes have become symbionts, it seems likely that mutualisms and various levels of protocooperation and parasitism are relatively fluid and dynamic conditions in lichens, both ecologically and evolutionarily. Indeed, interactions between fungus and alga in some lichen groups seem more like barely contained warfare than happy alliance. The bottom line is that lichen-forming fungi and algae, like other interacting forms of life, are basically self-interested opportunists.

# 6 | Cellular, physiological, and genetic traits

The preceding chapters have dealt primarily with PCM studies of macroscopic external features – organismal morphologies, behaviors, and lifestyles – that are often readily visible to the observer's naked eye. This chapter will illustrate how comparative phylogenetic analyses can likewise be conducted on microscopic internal attributes such as an organism's molecular makeup, cellular functions, physiology, genetic mechanisms, or its intragenomic "microbial associates" (including viruses and transposable elements). We will consider, for example, what PCM has revealed about the evolutionary-genetic underpinnings of sex determination, of eye development, of metazoan (multi-cellular animal) body plans, and of cellular mechanisms for DNA repair. We will consider how various fishes evolved antifreeze proteins, the capacity to produce electrical currents, and warm-bloodedness. And we will track the recent evolutionary history of the HIV viruses that cause AIDS. In truth, almost any subject related to cellular biology, ranging from biochemistry to medicine and epidemiology, can be informed to one degree or another by comparative phylogenetic analyses.

## Foregut fermentation

Most of the studies described in this book have employed molecular data as phylogenetic backdrops for interpreting the evolutionary histories of morphological or other organismal phenotypes. The general rationale, of course, is that, when species are assayed for hundreds or thousands of detailed genetic characters (as is typically the case with protein or DNA sequence information, for example), any widespread and intricate molecular similarities that might be observed are very unlikely to have arisen by convergent evolution, so they must instead reflect true phylogenetic descent. With secure phylogenetic foundations established from molecular analyses, any instances of convergence in organismal phenotypes (such as particular plumage features in birds, placental development in fishes, or viviparity in reptiles) then become much easier to identify and properly interpret.

This essay describes a case in which this logic has been applied in the opposite direction. In this inverted situation, an exceptional instance of evolutionary convergence at the molecular level has been deduced by reference to a phylogenetic framework securely established from traditional phenotypic (morphological and other) evidence. Convergent evolution at any level of biological organization is a powerful indicator of optimal (or at least recurrently favored) evolutionary design. When exquisite details of molecular structure or function can be firmly documented in unrelated lineages (and when horizontal gene transfers and retentions of ancestral conditions can be eliminated as potential explanations), the role of natural selection in repeatedly and independently shaping particular adaptations can become evident even at the levels of nucleic acids, proteins, or physiological operations.

Foregut fermentation is a phenotypic feature that clearly has evolved independently in diverse vertebrate lineages. In ruminants (such as the cow (*Bos taurus*)), in colobine monkeys (e.g. the Hanuman Langur (*Presbytis entellus*)), and also in a leaf-eating bird (the Hoatzin (*Opisthocomus hoazin*)), an anterior portion of the stomach has become modified to support fermentative bacteria that assist their hosts in digesting fibrous plant matter. Leaves and plant shoots are abundant in most environments, but they are also low in nutrition (per unit volume) and otherwise difficult to digest. Thus, the independent evolution of foregut fermentation by ruminants, langurs, and the Hoatzin (the only bird known to have such a capability) gave each of these animals the capacity to exploit an expanded smorgasbord of dietary options.

Scientists have discovered that these three evolutionary transitions to foregut fermentation also entailed convergent evolution at the sub-microscopic level of protein molecules. Lysozymes are a class of bacteriolytic (bacteria-destroying) enzymes found routinely in the tears, saliva, and other secretions of most animals. Their function is to kill invasive, potentially harmful microbes. In ruminants, langurs, and hoatzins, particular lysozymes have also been recruited to play a key role in foregut fermentation. In the highly acidic digestive systems of these animals, these lysozymes lyse (split open) the fermentative bacteria as they pass through the gut, thereby enabling the host to absorb and utilize valuable nutrients that had been assimilated by those bacteria. Many bacterial cells are killed in the process, but never mind: they had already copiously multiplied in the foregut, so vast colonies of their relatives carry on in a mutualistic room-and-board contract in which the host provides the bacteria with housing in return for their help with the overall food budget.

Environmental conditions in the digestive tract of each host tend to be quite unlike those in tears or other secretions where ancestral lysozymes originally

Hanuman Langur and domestic cow

operated. Accordingly, specific lysozymes that were later recruited to digestive services in the gut now display several novel biochemical features (e.g. low pH optima and resistance to protein-digesting enzymes) that enable them to function properly in their adopted tummy niche. These properties are shared by the gut lysozymes in ruminants, langurs, and hoatzins, and they therefore show how key aspects of protein function can sometimes evolve convergently in distant phylogenetic lineages. Even more remarkable are various fine details of protein structure that have been documented to underlie these functional convergences. In particular, in each vertebrate group displaying foregut fermentation, a unique combination of five amino acid replacements consistently distinguishes stomach lysozymes from their respective genetic homologs that continue to operate elsewhere in the host's body. Furthermore, phylogenctic analyses have also shown that significant accelerations in the molecular evolution of gut lysozymes accompanied each such recruitment of these enzymes to their new tasks in mammalian and avian stomachs.

In general, molecular evolution is primarily divergent rather than convergent. For example, if the broader genomes of ruminants, langurs, and hoatzins were to be compared, most homologous genes and their protein products would prove to have accumulated vast numbers of structural and functional differences over the long evolutionary times since these animals last shared common ancestors. Indeed, if molecular evolution were not predominantly divergent overall, molecular phylogenies would tend to contain more noise than historical signal and would be genealogically uninformative (which is patently untrue based on empirical experience). It is exactly for this reason that occasional cases of striking convergence in particular molecular features (as in the stomach lysozymes of vertebrate foreguts) are of special scientific interest. They provide compelling evidence that natural selection can, on occasion, promote recurring structural and functional themes in molecular (as well as in broader organismal-level) features.

### Snake venoms

Of the world's nearly 3000 extant species of snake, about 80% belong to the Colubroidea, a taxonomic superfamily that contains all of the known venomous taxa. Relatively few colubroid species produce powerful venoms, but those that do (such as coral snakes and rattlesnakes) often concoct incredibly potent brews. Snake toxins are secreted by special poison glands, and then, depending on the type of snake, are injected into victims by one or another delivery device such as enlarged front fangs, relatively undifferentiated teeth, or highly mobile fangs at the rear of the jaw.

Recent analyses based on DNA sequences and comparative morphology have improved scientific estimates of snake phylogeny and thus provided new historical perspectives on features associated with venomousness. For example, a majority (albeit not yet consensus) view is that poison-secreting glands arose near the base of the colubroid radiation, and that "evolutionary tinkering" (including multiple losses as well as elaborations of venom-inoculation devices) took place later.

The venoms themselves have recently become subjects of phylogenetic analysis. One discovery is that each category or family of poisons appears to have arisen when one or another standard body protein became extensively modified and evolutionarily recruited into the chemical arsenal of a particular snake lineage. Evidence for

Cottonmouth Snake

this notion has come from the detailed molecular analyses of modern snake venoms whose constituent proteins, although greatly altered, still tend to retain the basic scaffoldings (i.e. historical signatures) of the original body proteins from which they evolved. Distinct toxin families are numerous and diverse. They involve such biochemical classes as cystatins, lectins, natriuretic peptides, cysteine-rich secretory proteins, phospholipases, and others. Snake venoms in various species often contain multiple toxin types, which only serves to increase the overall potency of these poisonous potions. A species' total arsenal of toxins is sometimes referred to as its "snake-venom proteome."

Previously, the compositions of only select venoms from a few medically important snake species were well characterized, but recent molecular research has extended the analyses to several additional taxonomic lineages and venom types. Results from one such study, by Fry and Wüster (2004), are highlighted here. Figure 6.1 shows a composite phylogeny, based on several lines of evidence, for seven of the largest taxonomic groups of colubroid snakes. It also displays the interpretation by Fry and Wüster of where more than a dozen different types of toxins most likely originated in snake evolution. These provisional conclusions came from the authors' following observations and logic.

Fry and Wüster first compiled and compared sequence data on different types of snake toxin in extant species representing Elapidae and Viperidae, the two most

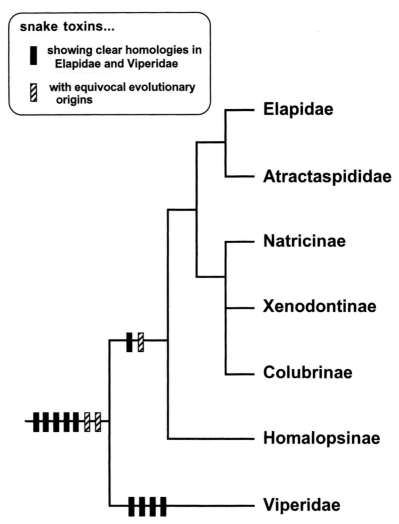

*Figure 6.1.* Inferred recruitment events of 13 different toxin types along the colubroid phylogenetic tree (after Fry and Wüster, 2004). Solid bars indicate toxins whose molecular make-ups provide strong support for evolutionary homologies between Elapidae and Viperidae; hatched bars refer to toxins whose evolutionary origins remain more equivocal. For the most part, only members of Elapidae and Viperidae were examined in this analysis, so the primary focus was on deep portions of the tree (thus, toxins that may have been uniquely recruited by derived lineages later in snake evolution would not appear on this phylogeny).

distant branches in the colubroid phylogenetic tree. Elapidae is comprised mostly of coral snakes, whereas Viperidae (the pit vipers) includes copperheads, cottonmouths, and rattlesnakes. Then, the authors reasoned that if even some extant snake species in these two evolutionary lines share specific toxins that themselves show a strong indication (in their detailed molecular structures) of being phylogenetically linked, then those toxin types must have originated prior to the evolutionary radiation of the Colubroidea. On the other hand, if a specific toxin type was confined to only one clade or subclade within the colubroid phylogeny, the possibility remains that it originated later, along that branch alone (or, alternatively, that it had been secondarily lost during evolution in the other colubroid clades).

The results, summarized in Fig. 6.1, provisionally indicated the following. First, the recruitment of at least five toxin classes into the snake-venom proteome apparently pre-dated the evolutionary diversification of colubroid snakes. Second, at least four toxin types probably arose later, near the base of the viperid clade. Finally, at least one class of toxins probably originated near the base of the viperid's sister clade (which includes Elapidae). Additional phylogenetic analyses and more comprehensive molecular characterizations will be needed to bring the full panorama of snake-venom evolution into sharper focus.

The data currently available also give strong hints about the venom proteome of the earliest colubroid snakes. Apparently, that original ancestor already had evolved complex and sophisticated brews of poisonous secretions whose functions may well have been similar to those of modern venomous snakes. As venoms from more snake lineages are examined and phylogenetic estimates are improved (for the snakes themselves as well as for particular toxins), a likely outcome may be that numerous toxins were also added, further modified, and lost (perhaps recurrently) on numerous occasions later in colubroid evolution. If so, the general pattern of venom evolution in snakes may parallel that of their poison-secreting glands: ancient basal origins, and a great deal of subsequent evolutionary tinkering.

## Antifreeze proteins in anti-tropical fishes

Marine vertebrates now living in polar regions must survive bone-chilling oceans where temperatures routinely reach −1.8 °C (the temperature below which seawater freezes). It was not always thus. Until about 15 million years ago, much of the Southern Ocean surrounding major landmasses was much balmier, but as Antarctica continued to separate from Australia and South America by plate tectonic movements, hydrographic changes resulted in a progressive reduction of ocean temperatures to the frigid Antarctic conditions of today. In the Northern Hemisphere, high-latitude oceans experienced dramatic cooling beginning about 2.5 million years ago with the advent of Pliocene and Pleistocene glaciations on massive continental scales.

Warm-blooded or endothermic polar birds (such as penguins) and mammals (such as seals and Arctic foxes) have evolved thick layers of feathers or fur, and subcutaneous fat, as insulation to help conserve body heat that is generated internally.

Giant Antarctic Icefish

Cold-blooded polar creatures such as fishes, by contrast, being at the relative mercy of external thermal regimes, have had to evolve other kinds of defenses against extreme cold. Otherwise, their body fluids and tissues would crystallize and freeze solid in these sub-zero waters. In the entire scientific literature, one of the hottest discoveries regarding convergent evolution at the molecular level involves the antifreeze glycoproteins (AFGPs) that are displayed alike by two phylogenetically and geographically distant groups of polar marine fishes.

Approximately 100 species of icefish (Notothenioidea, order Perciformes, super-order Acanthopterygii) constitute more than 90% of the Antarctic's piscine fauna in terms of species diversity and biomass. Most of the species examined to date possess distinctive glycoprotein molecules that inhibit the growth of ice crystals, depress the temperature at which the fish's body fluids freeze, and in general diminish each animal's susceptibility to death in much the same way that glycol-based antifreeze compounds help protect a car's operations during harsh winters. Each AFGP is composed of a disaccharide (a carbohydrate) linked to simple peptide repeat units encoded by various members in a large family of AFGP genes. Detailed

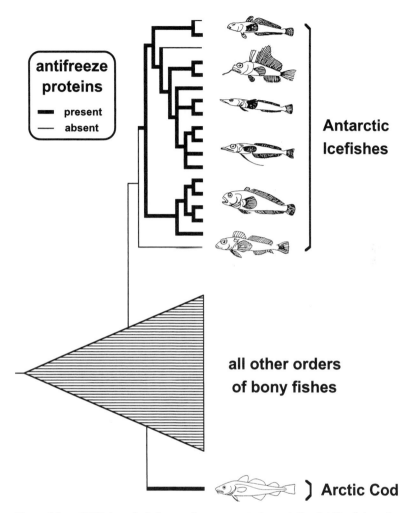

*Figure 6.2.* mtDNA-based phylogeny for representative notothenioid icefishes of the Antarctic (after Bargelloni *et al.*, 1994). Also shown is the known or suspected presence versus absence of glycoprotein antifreezes in these fishes (not all species have yet been examined), and in Arctic cod. As depicted, these Antarctic and Arctic fishes apparently trace to entirely separate evolutionary origins in the bony fish clade.

molecular characterizations have revealed that the AFGP genes in notothenioids probably evolved from a pancreatic trypsinogen-like ancestral locus.

On the other side of the world, several species of Arctic cod (Gadidae, order Gadiformes, superorder Paracanthopterygii) likewise possess AFGPs whose molecular structures (at the post-translational level) and freeze-protection functions are essentially identical to those of the Antarctic notothenioids. None the less, various lines of evidence indicate that the AFGP genes themselves had completely separate evolutionary geneses in cod and icefish. First, these two fish groups are distant both phylogenetically (as reflected by their membership in different taxonomic superorders) and geographically (they are now, and presumably always have been, confined to opposite poles of the planet). Indeed, morphologic, paleontologic, and paleoclimatic evidence all are consistent with the notion that these two fish lineages diverged well before antifreeze proteins were needed. Second, phylogenetic analyses indicate that the AFGP sequences in cod did not evolve from an ancestral trypsinogen gene (unlike the situation in icefish). Finally, several molecular details regarding AFGP genes (such as the precise boundaries between intron and exon sequences) differ between cod and icefish in ways that bespeak independent evolutionary origins.

So, the phylogenetic backdrop shows quite conclusively that AFGP molecules were gained at least once near the base of the Antarctic icefish clade, and again independently in Arctic codfish. Furthermore, these AFGPs have been lost, probably secondarily, in some icefish species (see Fig. 6.2) that inhabit more benign thermal latitudes just outside the sub-zero Antarctic zone. Thus, the mere presence versus absence of these remarkable biochemical adaptations reveals more about the thermal environments occupied by these fish than it does about the fishes' phylogenetic affinities. The same would of course be true of cars' glycol-based antifreezes, which are needed in cold climates regardless of the particular factories in which the automobiles were produced.

### Warm-bloodedness in fishes

Nearly all of the world's 30 000 teleost fish species are cold-blooded or ectothermic, meaning that their bodies typically remain within 1–2 degrees of ambient water temperatures. One reason is that all teleosts respire through gills with large surface areas. There, a close apposition of circulating blood with colder ambient water would normally promote the rapid dissipation of any body heat that these animals might otherwise generate by elevating the aerobic capacities in their muscles or other tissues. Nevertheless, several species of large oceanic predator in the order Scombroidei (billfish, tuna, and mackerel) do maintain appreciably elevated body temperatures by metabolic means. The evolution of endothermy by these fishes is especially remarkable because many scombroids live in frigid high-latitude oceans.

To illuminate where and how endothermy evolved within Scombroidei, Block *et al.* (1993) mapped the distributions of relevant physiological traits onto a phylogenetic tree (estimated from mtDNA sequences) for approximately 30 species of billfish, tuna, and mackerel (Fig. 6.3). The researchers thereby documented

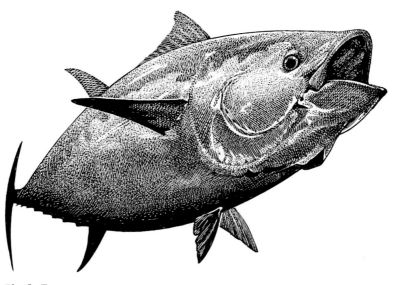

Bluefin Tuna

that endothermy in these fishes had at least three separate geneses, each entailing a unique suite of adaptations falling into one of two main physiological
categories.

Members of the tuna clade achieve endothermy in a manner roughly similar
to birds and mammals, i.e. by displaying exceptionally high metabolic rates and
reducing whole-body thermal conductance. The red aerobic muscle that makes
tuna powerful swimmers generates the majority of this metabolically derived heat,
much of which is retained within the body by way of vascular countercurrent heat
exchangers in the brain, muscles, and viscera. Heat retention is also facilitated
by the positioning of red muscle masses in a centralized body location close to
the vertebral column (unlike the lateral location of red muscles, just under the
skin, in most other teleosts). In contrast, all other scombroid species that generate
heat internally display cranial endothermy. In this more restricted version of the
elevated-temperature phenomenon, only the fish's brain or eyes are heated well
above ambient conditions.

Cranial endothermy itself appears to have had two separate physiological and
evolutionary etiologies. In all members of the billfish clade (swordfish, sailfish, and
marlins), a thermogenic organ composed of superior rectus muscle fibers near
the eye produces metabolic heat that is then retained locally by modified blood
vessels and by a countercurrent heat exchanger beneath the brain. In the one
and only known endothermic species of mackerel (*Gasterochisma melampus*), a
thermogenic region is similarly present in the head, but it is composed of lateral

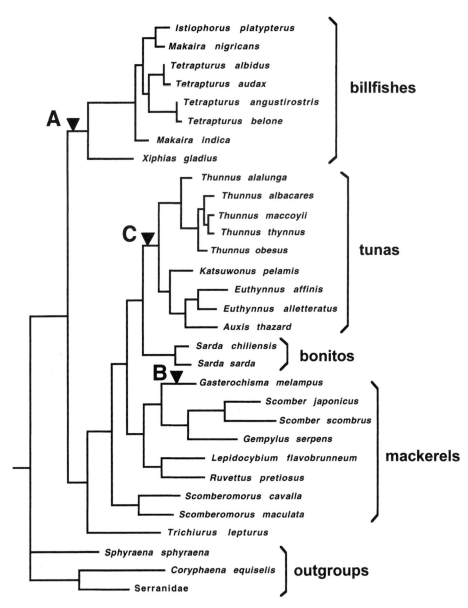

*Figure 6.3.* Molecular phylogeny for representative scombroid species, as based on mtDNA sequence data (after Block *et al.*, 1993). The letters A, B, and C denote the deduced evolutionary origins of three physiologically distinct forms of endothermy in these fishes (see text).

rectus muscle fibers that derive during ontogeny from cell types different from those underlying the billfish heater organ.

Thus, in scombroid fishes, endothermy appears to be triphyletic, cranial endothermy is diphyletic, and each of three distinctive physiological means of generating and maintaining heat is probably monophyletic (see Fig. 6.3). These varied phylogenetic outcomes arise from and illustrate an important distinction often made between evolutionary "attributes" and "characters" (as defined in their narrow rather than catholic versions). By strict definition, an attribute (such as endothermy) is a broad or composite summary of an organism's general pheno- type, whereas a character (such as cranial endothermy) is a more focused pheno- typic description. Obviously, these are notions along a continuum. Thus, cranial endothermy can also be considered a reducible attribute, as perhaps can each class of muscle fiber that underlies a thermogenic organ, and so on. The salient idea is that phylogenetic dissections of organismal features at multiple levels of inclusiveness along an attribute-to-character continuum can be collectively more enlightening about evolutionary origins and pathways than are phylogenetic analy- ses confined to any one hierarchical level alone.

The discovery that endothermy has had recurrent independent origins in the Scombroidei suggests that there has been strong selection in this taxonomic group for what is otherwise a costly and rare metabolic strategy in fishes. One possibility is that elevated body temperatures in tuna and their allies are simply associated with increased aerobic activity that has permitted improved locomotory performance in these top-echelon predators. Another possibility (not incompatible with the first) is that endothermy has been selectively favored because it enables scombroids to exploit a broader range of thermal regimes. For example, Swordfish (*Xiphias glad- ius*) often forage on squid at great oceanic depths where waters may be far colder than near the surface, and some scombroids such as the Bluefin Tuna (*Thunnus thynnus*) spend much of their lives in cold prey-rich waters beyond 45° latitudes north and south. Especially in such environments, it may be important for visu- ally oriented predators to warm key components of their nervous system (brain and eyes), and indeed such warming is a notable feature of all three lineages of endothermic scombroids.

Apart from an active lifestyle, another factor that probably preconditions a piscine lineage for endothermy is large body size (all else being equal), because only in large fish can deep tissues remain adequately insulated from direct envi- ronmental exposure. Indeed, elsewhere in fishes, endothermy is known to have evolved only in large sharks of the family Lamnidae (which includes the Great White Shark, *Carcharodon carcharias*).

### Electrical currents

Organisms living in aqueous environments generate mild electrical fields due to slight ionic differences between their body tissues and the surrounding water. Even slight muscular activities, such as heartbeats or respiratory movements, further amplify these electrical currents. In turn, various fishes in diverse taxonomic groups have evolved specialized receptor cells that can detect these naturally emitted electrical signals. When stimulated, these receptor cells release chemical neurotransmitters that then activate neurons that carry sensory impulses to the fish's brain. Many sharks and catfish, for example, employ their electroreceptive capabilities to help locate prey, and some nocturnal fishes use electrical currents to recognize and communicate with one another (see below).

Several piscine lineages have taken electrical engineering a step farther by evolving body organs that purposefully generate electric fields. For example, electric organs embedded in the lateral musculature of South America's Electric Eel (*Electrophorus electricus*; Electrophoridae) can discharge pulses of more than 500 volts at about 1 amp, which is ample to stun nearby fish and other prey items (and also to injure larger animals, including people). In Africa's electric catfishes (Malapteruridae), electric organs run the length of the body just under the skin. They are composed of muscle tissues arranged in a series, like plates in a car's battery, and they can generate a current of about 300 volts. Other fish that generate electrical currents to capture prey, ward off predators, or communicate with one another include stargazers (Uranoscopidae) whose electrical organs are derived from ocular muscles, ghost knifefishes (Apteronotidae) whose electricity comes from modified nerve fibers, and various species of electric rays (Torpedinidae) that have paired electric organs often in the head region or on the "wings." The fact that electric organs are of varied designs and are possessed by only a few disparate taxonomic groups of bony fishes (Actinopterygii) and cartilaginous fishes (Chondrichthyes) strongly suggests that this stunning adaptation had polyphyletic origins.

Another electrifying piscine group consists of approximately 200 species of elephantfish (Mormyridae), so named because some of these species display a long trunk-like proboscis. In these mostly nocturnal animals, native to central and southern Africa, electrical currents are generated by modified muscle tissue in the caudal peduncle region next to the tail. This organ produces an electrical field that surrounds the fish and, in conjunction with electroreceptors, enables the animals to detect obstacles and food even in dark or turbid waters. Elephantfish probably use their electric organs to communicate with other individuals in such behavioral contexts as courtship, mating, aggression, and spatial or social cohesion.

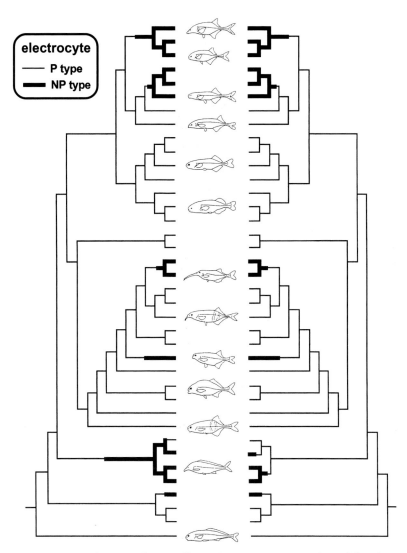

*Figure 6.4.* Two alternative but equally parsimonious reconstructions of electric organ cell types on a molecular phylogeny for 38 species of elephantfish (after Lavoué *et al.*, 2003). The phylogenetic tree was estimated from DNA sequences from the nuclear and mitochondrial genomes.

Molecular phylogenetic analyses have been conducted to illuminate the evolutionary histories of different types of electricity-generating cells (electrocytes) in these elephantfishes. Of special interest to neurophysiologists is a distinction between electrocytes in which the stalks of the electromotor neurons penetrate the cell layer (P type cells), versus those with non-penetrating stalks (NP types). In electric organ discharges (EODs), P cells are suspected to reduce the undesirable

DC (direct current) component. Otherwise, the DC currents might partly jam the fish's own receptor cells (perhaps making it harder for the fish to detect prey), or they might act as stronger homing beacons to the elephantfish's own predators (such as catfish). From these considerations, P cells would generally seem to be favored by natural selection.

Results from the PCM exercise are plotted in Figure 6.4, which shows two equally parsimonious interpretations (each involving seven evolutionary steps) of how P and NP electrocytes map onto a molecular phylogeny derived from DNA-level data. Under both scenarios, P cells were inferred to be the ancestral condition for Mormyridae, from which NP cells secondarily emerged on multiple evolutionary occasions. The first PCM scenario (left side of Fig. 6.4) involves six independent evolutionary switches from P cells to NP cells, and one reversion from NP cells to P cells. The second scenario (right side of Fig. 6.4) entails seven independent evolutionary changes from P cells to NP cells.

If P cells usually are favored by natural selection (for the reasons discussed above), then why have multiple evolutionary reversions to NP cells apparently taken place in elephantfishes? One intriguing hypothesis, raised by Carl Hopkins and colleagues, invokes sexual selection. NP cells are suspected to facilitate EODs with long-duration waveforms that may facilitate inter-individual communication. Hopkins' hypothesis is that sexual selection might override natural selection, and thereby give a net selective advantage to NP cells, in any elephantfish species in which females tend to select mates with long-duration EODs. The idea (which remains speculative at present) is that long-duration electrical discharges by elephantfish males, like long-duration songs by some songbird males, might signal to a female that her prospective mate is of exceptionally good health and/or high genetic quality. In other words, males that exude more sexual electricity may be at a tremendous advantage in the mating game.

### The Xs and Ys of sex determination

Humans and other mammals have a male-heterogametic system of chromosomal sex determination in which each male normally displays one X and one Y chromosome, each female carries two X chromosomes, and the sex of each offspring is determined by whether an X-bearing or Y-bearing sperm cell fertilizes a mother's X-bearing egg. The situation is exactly reversed in the female-heterogametic system of birds, in which each male carries two Z chromosomes (analogous to the mammalian X), each female displays one W chromosome (analogous to the mammalian Y) and one Z, and the sex of each offspring is determined by whether a Z-bearing sperm cell fertilizes a W-bearing or a Z-bearing egg.

These are just two of several ways in which sex is determined in various verte-brate groups. Some fish species, for example, lack recognizable sex chromosomes, but specific genes none the less channel an individual's early development into a male or a female pathway. Such genetic systems are termed genic rather than chromosomal (although sex-chromosomal systems also involve sex-determining genes with major developmental effects). Still other creatures show environmental (rather than genetically hard-wired) sex determination. In many turtles, for exam-ple, low and high egg-incubation temperatures trigger the production primarily of males and females, respectively.

All of the above refers to dioecious species, i.e. those with separate sexes, but the notion of directed differentiation into males or females can itself be rather fluid. Hermaphroditic fishes exist in which each individual may function both as a male (sperm producer) and a female (egg producer) either simultaneously or sequen-tially during its lifetime. At the other extreme, some lizard and fish species con-sist solely of females who reproduce clonally by virgin birth (see *Parthenogenesis*, Chapter 4). Among vertebrate animals, fish collectively show the greatest repertoire of sex-determining modes. The world's 25 000 living fish species include groups with mammalian-like X–Y systems, avian-like Z–W systems, non-chromosomal or genic systems of sex determination, parthenogenetic and hermaphroditic repro-duction, and taxa with various forms of environmental sex determination.

How and how often do evolutionary lineages switch from one mode of sex deter-mination to another? Theorists have puzzled over such questions and come up with several hypothetical models. For example, suppose that a lineage initially con-sists of hermaphroditic (cosexual) individuals whose male and female sex organs are determined by two separate genes ($M$ and $F$, respectively), and suppose then that a male-sterility mutation ($M^s$) arises from the normal male-fertile ($M^f$) gene, thereby transforming some cosexuals into females. Males should then be relatively more favored in the population because in effect they have become scarcer. So, any female-sterility mutation ($F^s$) might also increase in frequency under natural selec-tion, and together with the $M^s$ gene perhaps eventually convert a hermaphroditic population to a dioecious one. Theoreticians have further shown that such a hypo-thetical process would be more likely if the $M$ and $F$ genes became tightly linked on a chromosome such that favored combinations of genes ($M^f / F^s$ yielding fertile males, and $M^s / F^f$ yielding fertile females) would be evolutionarily perpetuated and eventually elaborated into classic Y and X chromosomes. Genetic details of such theoretical models are beyond the current scope, but the point is that scenarios can be envisioned in which evolutionary interconversions between hermaphroditic and dioecious populations, and between genic and chromosomal systems of sex determination, are plausible.

Such evolutionary switches are unlikely to be observed during any researcher's lifetime, but their past occurrences can be unveiled by PCM. One such study used parsimony and likelihood approaches (see Appendix) to plot the inferred evolutionary histories of alternative sex-determining mechanisms in bony fishes, using molecular phylogenies as backdrops. Figure 6.5 provides two such plots that illustrate the study's most general finding, namely that sex determination in fishes is an extremely labile evolutionary trait. Thus, numerous independent evolutionary transitions have occurred between, for example, hermaphroditism and dioecy, and between different genetic systems (e.g. X–Y and Z–W) of sex determination. These findings also clearly indicate that each mode of sex determination is polyphyletic in bony fishes. Indeed, this evolutionary lability was generally so pronounced as to preclude definitive statements about the exact ancestral states of sex determination for most clades.

The PCM analyses also highlighted a remarkable distinction between fish on the one hand, and mammals and birds on the other, regarding the evolutionary flexibility of sex-determining mechanisms. Whereas mammals and birds have steadfastly retained their respective X–Y and Z–W systems for many tens of millions of years, many piscine lineages over comparable evolutionary timescales have interconverted readily between these and additional sex-determination systems. Reasons for this disparity remain conjectural but probably have to do with greater developmental plasticities in fishes. For example, unlike the situation in birds or mammals, testes and ovaries in fishes can rather flexibly differentiate at almost any life stage within an individual. Many fish are sequential hermaphrodites, beginning their reproductive life as one sex and then later switching to the other. Presumably, such developmental flexibility has opened wider windows of opportunity for environmental pressures such as social status, ecological conditions, or population composition to play proximate roles in sexual expression, and also, via the selection pressures they impose, to promote frequent evolutionary changes among underlying sex-determining mechanisms.

### The eyes have it

Biblical creationists, and others who are philosophically opposed to the notion of evolution, often tout the vertebrate eye as an example of a complex feature that only the hands of a conscious creator (a sentient God) could have engineered. How, they rightly ask, could an organ so perfect have arisen by unconscious natural processes? Charles Darwin himself was acutely aware of the challenge to natural selection posed by elaborate attributes such as the eye, stating in *The Origin of Species* that "If it could be demonstrated that any complex organ existed, which could not possibly

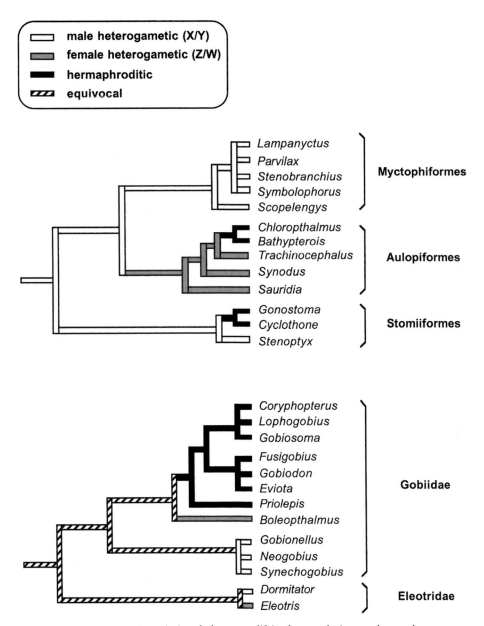

*Figure 6.5.* Two piscine clades exemplifying how evolutionary changes between alternative sex-determination mechanisms must have occurred quite frequently in fishes (after Mank *et al.*, 2005). In these cases (and others like them), the phylogeny was estimated primarily from molecular genetic data, and the PCM reconstructions were based on empirical observations of sex-determination modes in extant specics.

Blue-ringed Octopus (mollusk) and Bushbaby (primate)

have been formed by numerous, successive, slight modifications, my theory would absolutely break down." Darwin also noted, however, that "if numerous gradations from a simple and imperfect eye to one complex and perfect can be shown to exist . . . then the difficulty of believing that a perfect and complex eye could be formed by natural selection, though insuperable by our imagination, should not be considered as subversive of the theory."

Darwin knew of the great diversity of animal eyes, which range from the relatively simple photoreceptor tissues of some flatworms and bivalve mollusks, to the multi-faceted compound eyes of insects, to the sophisticated eyes of octopuses and the camera-lens eyes of vertebrates. This high anatomical diversity is a double-edged sword for evolutionary explanations. On the one hand, it implies that photorecep-tor organs with a wide range of sophistication probably have adaptive utility, thus adding plausibility to Darwin's suggestion that complex eyes evolved from simpler precursors in stepwise fashion, with each successive step beneficial to its bear-ers. On the other hand, the high morphological diversity of photoreceptor organs could be interpreted to indicate polyphyletic origins for light sensing, in which case evolutionary explanations would have to account for multiple geneses of eyes (rather than just one). In an influential paper, Salvini-Plawen and Mayr (1961) used anatomical evidence to claim that photoreceptor organs evolved independently at least 40–65 times in different animal lineages. Indeed, the notion that eyes are polyphyletic has been standard wisdom in evolutionary biology for at least the past 100 years.

For this reason, great interest attended a recent scientific claim by Walter Gehring (2000): "the dogma of a polyphyletic evolutionary origin of the eye has to be abandoned." Gehring was referring to a major molecular discovery from his laboratory that a shared genetic groundwork underlies the outward diversity of animal eyes. In creatures as different as mammals, amphibians, fish, sea squirts, sea urchins, squid, nematodes, ribbonworms, and flatworms, a gene known as *Pax6* acts as a master control element necessary for eye morphogenesis. In all of these creatures (and probably many others less well studied to date), *Pax6* encodes a transcription factor that initiates a developmental cascade in which legions of genes that serve in the production of light-sensing organs are activated. As further genetic support for a monophyletic (as opposed to polyphyletic) mechanistic basis for the eye, Gehring and Ikeo (1999) note that all seeing metazoans also share the same visual pigment, rhodopsin.

The structural and functional universality of *Pax6* in eye morphogenesis is supported by at least two types of observation. First, amino acid sequences in the protein products of this gene, in creatures ranging from flatworms to mammals, all group together in a phylogenetic tree (Fig. 6.6), thus suggesting descent with modification from an ancient protein in a common ancestor. Second, experimental studies by Gehring's group have shown that various forms of *Pax6* remain functionally effective even when swapped between highly divergent organisms. For example, when *Pax6* sequences from humans, sea squirts, or squid are artificially transferred (by using recombinant DNA techniques) to mutant strains of *Drosophila* that lack the gene, these foreign loci successfully induce the formation of eyes in these mutants. Such substitutibility clearly indicates that *Pax6* exerts similar master control over eye morphogenesis in all sorts of vertebrate and invertebrate animals.

Do Gehring's findings truly overturn traditional wisdom that animal eyes are polyphyletic? In my opinion, the answer is "yes and no." It does now appear to be true that a key genetic component of eye formation is shared, by virtue of common ancestry, among creatures with otherwise highly diverse types of photoreceptor organ and eye. In this sense, homology (similarity by descent) has been demonstrated. But, by Gehring's own estimates, more than two thousand genes altogether contribute to the developmental pathway of a typical metazoan eye. It remains to be seen how many of these loci prove to be homologous in highly divergent creatures. So, the broader point (also emphasized by Gehring) is that homology need not be an all-or-nothing phenomenon. Some genetic or structural components of various eyes (for example) may be homologous and monophyletic whereas other

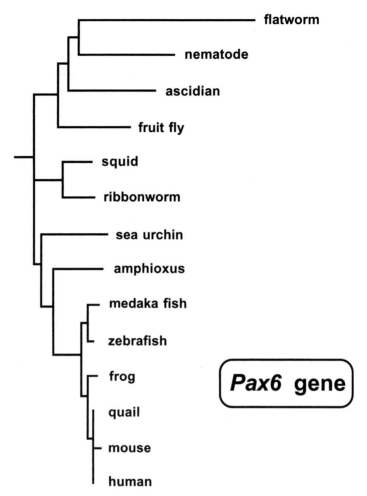

*Figure 6.6.* Phylogenetic tree estimated for *Pax6* genes from various metazoans (after Gehring and Ikeo, 1999). Branch lengths are proportional to numbers of amino acid substitutions.

components of those same features may prove to be analogous and polyphyletic (i.e. show similarities because of convergent evolution from different ancestral states).

### Two types of body

With regard to general body plans, two ancient lineages of metazoans are traditionally recognized: the Radiata, with radial symmetry (like circles or cylinders);

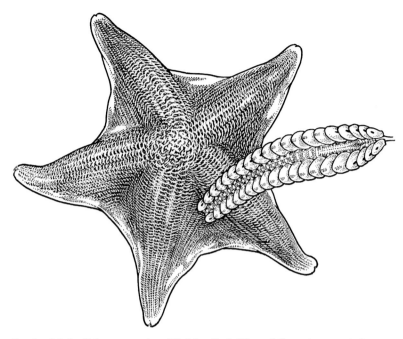

Bat Starfish (radial symmetry) and Pulchra Scale Worm (bilateral symmetry)

and the Bilateria, with lateral symmetry along the body axis (in which one end of the animal is consistently directed forward and a ventral surface is consistently oriented toward the substratum). More than 99% of living metazoans belong to the Bilateria lineage, which from fossil and other evidence arose more than 500 million years ago. Included in this clade are most or all species in such diverse taxonomic groups as Platyhelminthes (flatworms), Nematoda (roundworms), Annelida (segmented worms), Mollusca (snails, bivalves, and allies), Arthropoda (insects and their relatives), and Vertebrata (animals with backbones). The Radiata lineage, which is at least as ancient as the Bilateria, is today represented primarily by Cnidaria (e.g. anemones, hydras, and jellyfishes).

Actually, a few members of the Bilateria lineage, such as sea urchins and starfishes in the phylum Echinodermata, display radial symmetry, but a scientific consensus seems to be that these instances reflect the secondary re-evolution of this body plan well within the Bilateria lineage. Conversely, various members of the Radiata clade, such as some corals and sea anemones (in the phylum Cnidaria), actually show bilateral symmetry. The evolutionary origins of this condition are less clear, and two competing hypotheses have been put forward (Fig. 6.7): (A) the bilateral body plan pre-dated the phylogenetic split between Bilateria and Radiata, in which case bilateral symmetry in some cnidarians could be a retained ancestral condition;

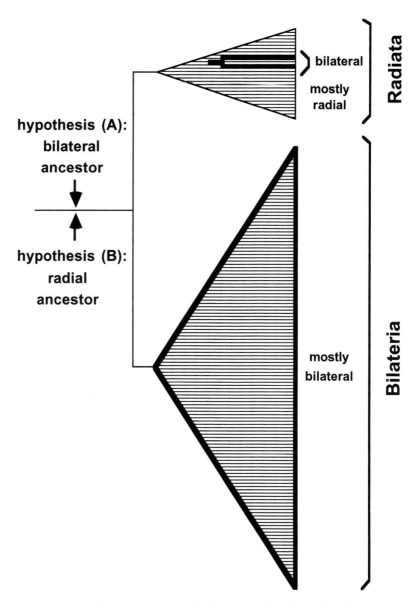

*Figure 6.7.* Two alternative scenarios for the ancestral condition of body ground plans prior to the evolutionary separation of Bilateria and Radiata lineages (see text). This depiction shows how some bilateral lineages are embedded within the otherwise mostly radial Radiata clade.

or (B) the radial body plan pre-dated the phylogenetic separation of Bilateria and Radiata, in which case some corals and sea anemones evolved bilateral symmetry independently of the Bilateria. Under debate is whether the initial ancestral condition for metazoans was bilateral symmetry (hypothesis A) or radial symmetry (hypothesis B), and the basic question can be phrased as follows. Does the presence of a bilateral body plan in the Bilateria and Radiata clades reflect homology (via retention of the ancestral state), or analogy (via evolutionary convergence, i.e. homoplasy)?

One powerful and general method to help distinguish homology from analogy is to examine finer details of a broader condition (such as a particular body plan) that is shared by two or more lineages, the rationale being that intricate similarities are unlikely to have arisen by convergent evolution and, thus, probably reflect true phylogenetic descent. One such recent analysis involved dissecting the underlying molecular genetic basis of bilateral symmetry (Finnerty *et al.*, 2004) in species representing Bilateria and Radiata. The following is a much abbreviated account of that and associated work.

Homeotic loci (see also *Snails' shell shapes*, Chapter 2) are genes with profound morphotypic influences during individual development (i.e. ontogeny). For example, more than half a dozen *Hox* genes in dipteran insects (from the Greek word *dipteros*, meaning two-winged) determine an animal's body-segment identities. Particular mutations (usually harmful to their bearers) at these loci can have remarkable consequences, such as developmentally transforming a standard fruitfly with two wings into a four-winged fruitfly, or transmogrifying what normally would be a fruitfly's antenna into a leg. Perhaps not surprisingly, it turns out that *Hox* genes also orchestrate bilateral development in Bilateria, by exerting regulatory control early in life on an animal's dorsal–ventral (DV) and anterior–posterior (AP) body axes. This has been understood for more than a decade.

What Finnerty and colleagues added to previous knowledge is a discovery that particular well-known *Hox* genes in Bilateria also appear to govern developmental body axes in surveyed bilateral species within the Radiata lineage, and in much the same operational way. For example, in both Bilateria and the bilateral Radiata species, a *Hox* gene by the name of *decapentaplegic* (*dpp*) is expressed asymmetrically along the developing DV body axis. Other such *Hox* genes are likewise expressed and seem to function similarly in initiating the development of body form along the AP axis.

The authors interpreted these findings as strong evidence for long-term evolutionary retention of homologous genes underlying bilateralism. This adds considerable support for hypothesis A, namely that bilateralism (or at least the presence of some of its key genetic elements) was an ancestral state that originated prior to any Radiata–Bilateria split.

## The phylogenomics of DNA repair

In 2001, a massive research endeavor known as the Human Genome Project came to fruition when the human genome, it all its three-billion-base-pair glory, was fully sequenced. Descriptions of the complete DNA sequences of several other vertebrate species, such as a mouse, chicken, and a fish, soon followed. Today, with further technological improvements, it has become almost routine for scientists to sequence entire genomes, especially of relatively simple creatures such as bacteria, whose full DNA complement is usually about 1000 times smaller than that of a typical vertebrate species. Altogether, complete genomic sequences are now available for more than 100 microbial and other taxa, and the list is growing rapidly. Science clearly has entered an era of mass genomics. According to Eisen and Hanawalt (1999), it has also entered a "phylogenomic era" in which the results of molecular genomic analyses can and should be interpreted in a phylogenetic context.

These authors illustrated their concept of phylogenomics by conducting an evolutionary analysis of DNA repair genes. Because DNA is continually subject to damages from mutagenic agents in the environment, and from occasional chemical errors arising during its own replication, the repair of genetic material is a fundamental challenge for all forms of life. Over geological time, species have responded by evolving a wide variety of cellular mechanisms for DNA repair. In one species or another, scientists have discovered repair pathways that can chemically mend just about any type of DNA abnormality: single- and double-strand breaks, intra- and inter-strand cross-links, chemically modified bases, base-pair mismatches, and others. Some of the DNA repair pathways are simple and involve single enzymes, whereas others are complex and involve as many as dozens of enzymes working in concert. Some of the pathways have single functions, whereas others play roles in diverse cellular processes. Some DNA repair genes seem to have unique duties, whereas others functionally overlap and thereby give partial redundancies to some aspects of DNA repair.

Eisen and Hanawalt tackled this diversity of DNA repair mechanisms by first computer-searching available genomic databases for the presence versus absence of dozens of different repair genes in a variety of taxa. The computer search involved full-genome scans of 14 microbial species representing the two primary evolutionary domains of prokaryotic life (Bacteria and Archaea), plus two species (yeast and humans) representing Eucarya (a third major evolutionary lineage whose species,

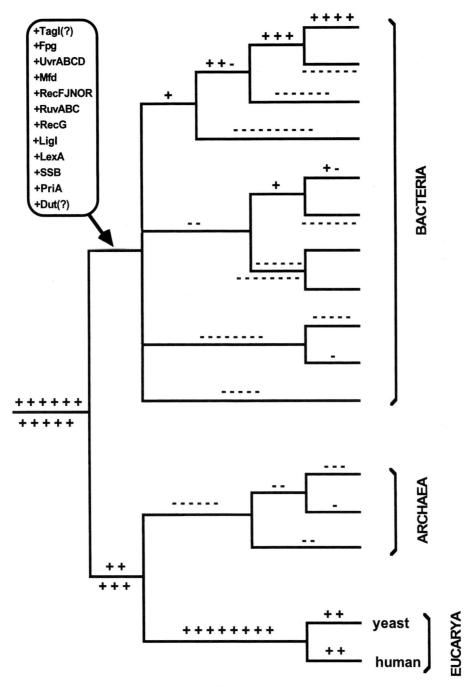

*Figure 6.8.* Cladogram for the three major evolutionary domains of life (Bacteria, Archaea, and Eucarya) onto which have been plotted inferred gains (+) and losses (−) of various genes involved in DNA repair (after Eisen and Hanawalt, 1999). Each gain or loss refers to a precisely characterized gene, but for simplicity (and for reasons of space), only one such set of loci (that listed along the branch leading to the Bacteria) is explicitly named here.

unlike prokaryotes, house their chromosomes in a membrane-bound cell nucleus). Then, using a molecular phylogenetic tree as backdrop, the authors used the logic of PCM to infer the evolutionary histories of gain and loss for each gene whose protein product was understood to play a role in the reparation of DNA damages.

This phylogenomic exercise (only the cursory results of which are summarized in Fig. 6.8) illuminated several points about the evolutionary histories of DNA repair mechanisms. First, different forms of life have quite different suites of repair genes and repair capabilities. Although some DNA repair loci (most notably RecA, which is involved in DNA recombination) proved to be ubiquitous or nearly so among the species surveyed, most repair genes were confined to particular branches of the phylogenetic tree. Indeed, some repair loci appeared to be of recent origin, indicating that the evolution of DNA repair capabilities is an ongoing process. Second, the phylogenetic distributions of DNA repair loci appeared to be due to evolutionary gains and losses of particular genes and gene pathways along specifiable branches in the phylogenetic tree. For example, the dozen genes explicitly listed in Fig. 6.8 most likely arose on the evolutionary line leading to Bacteria. Third, the phylogenomic analyses permitted conclusions about how some of the genetic changes underlying DNA repair mechanistically occurred. For example, detailed sequence analyses revealed that several gene-duplication events probably contributed to the proliferation of DNA repair loci at the basal root of the overall tree and also at the root of the Eucarya. Finally, the phylogenomic analyses helped to uncover some instances in which similar functional pathways of DNA repair probably had entirely separate evolutionary origins. The best-documented example of this phenomenon involved NER (nucleotide excision repair), which involves different suites of genes and cellular mechanisms in bacterial versus eukaryotic systems.

To a large degree, phylogenomic exercises of this sort are really just further examples of PCM, albeit at comprehensive genomic scales. Like other PCM endeavors, their value lies in fostering enriched evolutionary understandings of biological phenomena (in this case, the genetics of DNA repair) by adding historical perspectives to what otherwise would be merely contemporary accounts of functional processes in living species.

### Roving nucleic acids

Not all organismal phylogenies are 100% tree-like, i.e. strictly branched and hierarchical. Instead, vertical branches in some phylogenetic trees have been interconnected to varying degrees by various mechanisms of lateral or horizontal genetic transfer (LGT or HGT) that postdated the relevant nodes. (Note: vertical connections in an evolutionary tree conventionally refer to parent-to-offspring genetic transmission across the generations, whereas horizontal connections refer to lateral movements of genes between otherwise reproductively isolated lineages. However, because all phylogenies depicted in this book are rotated 90° relative to an upright tree, any LGT event would appear here as a vertical line secondarily connecting two left-to-right lineages.) If LGTs between species were frequent during the evolution of a particular taxonomic group, a proper historical representation could come to resemble, in the extreme, a reticulate or anastomotic genetic web more so than a traditional phylogenetic tree.

Introgressive hybridization is one way that genes sometimes transfer between closely related species. Many congeneric fishes, for example, hybridize at least occasionally in nature, and their resulting progeny, if viable and fertile, may backcross to one or the other parental species and thereby serve as a biological bridge for interspecific gene movement (i.e. introgression). Such instances are empirically well documented in numerous groups of fishes and other vertebrates, invertebrates, and plants. Gene transfer via hybridization is not strictly horizontal because vertical (parent-to-offspring) transmission across successive generations is involved. Nevertheless, when plotted onto a longer-term phylogeny, each occurrence of introgressive gene movement would appear as an instantaneous lateral genetic exchange between nearby tree branches.

Another route for LGT appears to be via transposable elements, or "jumping genes." These mobile elements (small pieces of DNA that in effect can hop from one chromosomal site to another) usually confine their activities to within a cell lineage, but empirical evidence suggests that they sometimes leap across species' boundaries as well. A case in point involves gypsy elements in *Drosophila* fruitflies. Each gypsy element is a viral-like piece of DNA about 7500 nucleotide-pairs long that encodes several proteins responsible for its own replication and infectivity. Each gypsy element normally resides passively at its ancestral homesite in the genome and is transmitted vertically across the generations, just like any ordinary gene. However, every once in a while, a copy of gypsy colonizes (via an RNA intermediate) a new chromosomal location. Rarely, one of these retrotransposable elements even hops into a different species. Such movements seem to be genuinely horizontal because, presumably, no vertical parent-to-offspring transmission is involved. Instead, gypsy and various other categories of jumping genes are thought to move between species either autonomously or by hitching rides on other biological vectors such as viruses, bacteria, or parasitic mites. Occasionally, some genes of the host species may accompany these mobile vectors on their travels.

Figure 6.9 summarizes several reported instances in which gypsy elements appear to have leaped species boundaries during *Drosophila* evolution. With respect to the mobile elements *per se*, these reticulation events partially convert

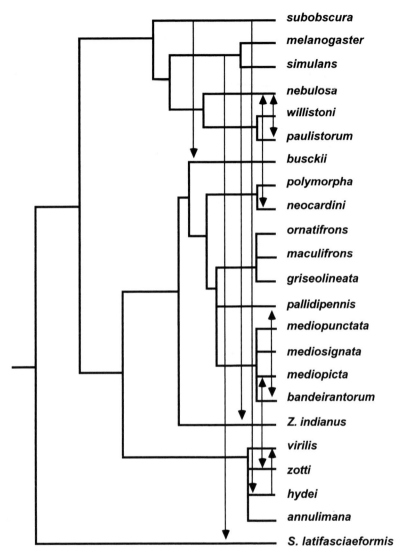

*Figure 6.9.* Evolutionary scenario for 23 species of *Drosophila* and related dipteran flies (after Herédia *et al.,* 2004). Thick lines, the species phylogeny based on a consensus of extensive molecular genomic evidence; thin lines with arrows, inferred instances in which gypsy elements apparently moved laterally between non-adjacent branches in the tree. These LGT events were estimated to have occurred between about 1.2 and 3.4 million years ago.

an otherwise straightforward phylogenetic tree into a more complicated network of historical connections. None the less, the overall basic tree structure for these species remains evident in extensive molecular phylogenetic analyses of many other components of the *Drosophila* genome. Indeed, it is such blatant discrepancies or incongruities between the nucleotide sequence phylogeny of particular gypsy elements on the one hand, and the consensus genomic phylogeny of the host species on the other, that provided the initial (and still strongest) evidence for occasional instances of lateral gypsy transfer during *Drosophila* evolution.

A third and even more remarkable form of reticulate evolution involves genomic mergers, the most famous example of which was an ancient marriage of microbial prokaryotes that led to the emergence of the first eukaryotic (nucleated) cells more than two billion years ago. At that time, one type of bacterium, bearing precursors of the mitochondrial genome, joined forces with another distinct type of bacterium, bearing precursors of many genes that were to become housed in the eukaryotic cell nucleus. Even today, a phylogenetic footprint of this event can be seen as the closer resemblance of some mitochondrial genes to those of modern Bacteria than to comparable genes housed in the nuclei of modern Eucarya (see *The phylogenomics of DNA repair*, above, for a description of life's basal phylogenetic domains). In plants, another such "endosymbiotic" merger eventuated in cells that carry a chloroplast genome in addition to those housed in the nucleus and mitochondrion. Many scientists now believe that several if not many genomic mergers of these sorts occurred early in the earth's history, and that the early Tree of Life was really more like an anastomotic bush.

All types of LGT, including those mentioned above, complicate phylogenetic analyses because they violate the usual assumptions of tree reconstruction. Interestingly, however, critical documentations of LGT almost always rely in large part on comparative phylogenetic analyses, because LGT events (except perhaps those mediated by introgressive hybridization) are relatively uncommon in evolution (compared with vertical transmission) and therefore are unlikely to be observed directly. Normally, each LGT is first identified, provisionally, as a gross topological disparity between the phylogenetic tree of a specific piece of DNA and the consensus organismal phylogeny. Follow-up research must then be conducted to eliminate competing hypotheses for the apparent discrepancy. The point is that, even in the context of studying LGT phenomena, where phylogenetic approaches might seem at face value to be least suitable, PCM analysis none the less remains an indispensable tool for drawing proper inferences about evolutionary processes.

### Host-to-parasite gene transfer

As described in the preceding section, not all DNA transmission during evolution is strictly vertical, i.e. from parents to their progeny. Occasionally, small pieces of genetic material move horizontally or laterally between species. The empirical evidence for such LGTs usually comes from a gross disparity between an overall species phylogeny (as evidenced by most of the genetic information within the relevant lineages) and the genealogy for the specific segment of DNA that presumably has been laterally transferred. In other words, the topology in a phylogenetic tree for the renegade piece of DNA may stand out like a sore thumb from the consensus phylogeny of the recipient genome in which it is currently housed.

In most instances, the precise mechanism underlying a given LGT event is unknown, but sometimes the biological circumstances point to a probable route of lateral movement. A case in point involves the remarkable "endophytic holoparasitic" association between parasitic plants in the family Rafflesiaceae and their host plants in the family Vitaceae. An endophyte is a plant that lives within another plant. A holoparasite is an obligate parasite, i.e. one that cannot live apart from its host. At some time in the not-too-distant evolutionary past, while physically locked in such an intimate endophytic embrace, it now appears likely that a small piece of DNA somehow transferred laterally from a vitacean host lineage to became incorporated into the mitochondrial genome of its rafflesiacean holoparasite.

Plants in the family Rafflesiaceae (order Malpighiales) are best known for their unusual morphology and lifestyle. Rafflesiaceans are physically housed in vitacean plants of the genus *Tetrastigma*; being nutritionally dependent upon their hosts, they lack the leaves, stems, and roots that are characteristic of free-living plants. Despite this extreme reduction in vegetative body parts, these holoparasites have unmistakable flowers, the largest and arguably the most outlandish in the world. These grotesque flowers, up to a meter across and growing well outside the host plants, mimic rotting flesh and thereby serve as an enticement to carrion flies that pollinate the flowers.

Morphological as well as molecular genetic evidence indicates that extant species of parasitic Rafflesiaceae are phylogenetically embedded within the Malpighiales, a huge clade of flowering plants comprising 27 taxonomic families and about a dozen orders. Thus, almost certainly, the Rafflesiaceae are not at all closely related to their vitacean hosts, which instead appear as basal dicotyledons (plants whose embryos have two or more leaves) in the phylogenetic tree of Angiosperms (flowering plants). These presumed relationships, in relevant outline, are summarized in Fig. 6.10. They are supported by phylogenetic analyses of DNA sequence data

*Rafflesia* flower

from both nuclear and mitochondrial genes, as well as by the fact that species in the vitacean clade are characterized by several distinctive synapomorphic features (such as seeds with cordlike seams, stamens opposite the petals, and characteristic chloroplast structures) that are not possessed by the rafflesiaceans.

Thus, it came as quite a surprise to the botanical community when Davis and Wurdack (2004) reported that DNA sequences from a single mitochondrial gene (*nad1*) phylogenetically grouped the parasitic rafflesiaceans with their vitacean hosts rather than with their supposed Malpighialean cousins. Indeed, if *nad1* had been the only guide to organismal phylogeny, then the parasite and host would seem to be extremely closely related, but this would flatly contradict the compelling phylogenetic evidence cited above that had come from DNA sequence data at other surveyed loci and from comparative morphology. Thus, *nad1* seemed to be a gross phylogenetic anomaly. The authors concluded that this locus must have been laterally transferred from host to parasite (see the arrow in Fig. 6.10) long after the relevant ancestral lineages had phylogenetically separated from one another in the far more distant evolutionary past.

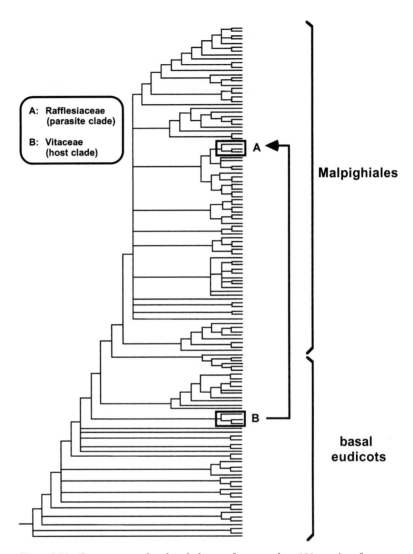

*Figure 6.10.*   Consensus molecular phylogeny for more than 130 species of
flowering plant (representing Malpighiales and various basal dicot lineages) based
on DNA sequences from several surveyed genes in the nuclear and mitochondrial
genomes (after Davis and Wurdack, 2004). The arrow shows the suspected lateral
transfer of the *nad1* gene from a vitacean host to its rafflesiace parasite (see text).

   Although this conclusion remains inferential, based on indirect (phylogenetic)
rather than direct mechanistic evidence, it does strongly suggest that small pieces of
DNA can sometimes be successfully exchanged between sexually isolated and phylogenetically divergent species living in close physical association (perhaps even
without the need for a mediating vector such as a virus or bacterium). This raises

the specter that further molecular phylogenetic examinations of parasites and their hosts (as well as participants in other types of symbiotic and non-symbiotic associations) may uncover many additional such examples of reticulate evolution. Indeed, recent studies (see examples listed in References) are beginning to paint phylogenetic pictures in which anastomotic connections attributable to LGT events may be far more common than formerly imagined between many otherwise divergent plant (and animal) lineages.

### Tracking the AIDS virus

Some organisms evolve so rapidly at the molecular level as to permit analyses of their genealogical relationships across recent years or even months. The best examples are retroviruses, infective and often disease-causing agents whose nucleic acids (RNAs in this case) mutate and diverge as much as 1 000 000 times faster than typical DNA molecules in the genomes of most other species. This exceptional pace of molecular evolution gives populations of RNA viruses tremendous adaptive potential, often including a proclivity to evolve genetic resistance to medical vaccines or to more natural countermeasures by the host. It also permits scientists to monitor evolutionary changes in retroviruses directly, in contemporary time.

Findings from such genealogical analyses can be important in medicine and epidemiology, as illustrated by studies of HIVs (human immunodeficiency viruses) that cause AIDS (acquired immune deficiency syndrome). These retroviruses come in two distinct classes (HIV-1 and HIV-2) whose ancestors apparently entered the human species early in the twentieth century in the form of simian immunodeficiency viruses (SIVs), which infect wild primates in Africa. Several historical details about the geographic sources and dates of these cross-species transfers, and the subsequent evolution of HIVs and the global spread of the AIDS pandemic within *Homo sapiens*, have been illuminated by phylogenetic reconstructions based on the retroviruses' nucleotide sequences.

With respect to origins, the molecular phylogeny summarized in Fig. 6.11A indicates that HIV-1 and HIV-2 probably evolved from the SIVs of Common Chimpanzees (*Pan troglodytes*) and Sooty Mangabeys (*Cercocebus atys*), respectively, shortly after these simian viruses somehow jumped into humans, perhaps on more than one occasion each. Much debate (mostly unresolved) has centered on how these retroviral transfers may have taken place, with hypotheses ranging from intimate human–simian tissue contact (e.g. when chimpanzees and monkeys were butchered for food), to the possibility of laboratory contamination of vaccines used to fight polio epidemics in Africa. The molecular phylogenies have not shed direct light on such causal mechanisms of transfer, but they have demonstrated

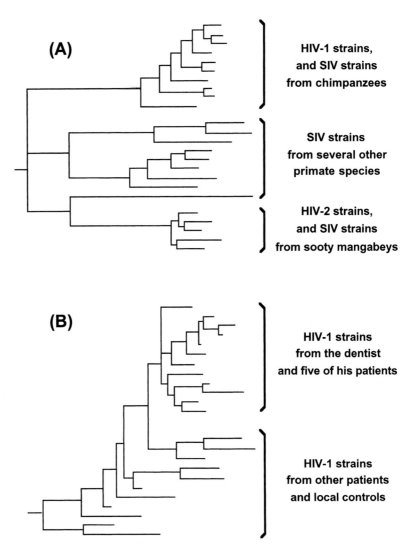

*Figure 6.11.* (A) Molecular phylogeny showing the genealogical relationships of human HIV-1 and HIV-2 viruses to SIV strains from various other primate species (after Hahn *et al.*, 2000). The SIVs included in this analysis came from the Common Chimpanzee, Sooty Mangabey and another unspecified mangabey species, several species of Guenon monkey (genus *Cercopithecus*), and the Mandrill (*Mandrillus sphinx*). (B) Molecular phylogeny for HIV-1 strains isolated from a Florida dentist (Dr Acer), several of his patients, and several local controls, i.e. other HIV-infected individuals from the same community (after Ou *et al.*, 1992). (Note: the two graphs are not drawn to the same scale, so branch lengths cannot be compared directly.)

unequivocally that the viruses causing human AIDS trace in recent evolution to more than one primate source.

With respect to tracking the subsequent global spread of AIDS, molecular phylogenetic appraisals have worked out many additional details, including in some cases precise estimates of colonization dates. For example, according to calculations by Korber *et al.* (2000) based on observed rates of viral sequence divergence, the entire HIV-1 pandemic traces to a common ancestral retroviral sequence dating to the early 1930s. Similar phylogenetic analyses of HIV gene sequences following AIDS outbreaks in the Americas led Li *et al.* (1988) to conclude that the viruses probably reached Haiti (from Africa) during the period 1969–75, and then spread onward to Florida and elsewhere in the United States by 1975–9. The most remarkable aspects of these phylogenetic reconstructions are the short timeframes involved. Molecular clocks for these HIV viruses tick so fast that large numbers of detectable nucleotide substitutions accumulate in a matter of years and decades rather than requiring millennia or eons.

Molecular phylogenetic analyses of HIV sequences also have found forensic applications in criminal justice. A famous case in Florida involved an HIV-positive dentist (Dr David Acer) and seven of his patients who became infected with HIV-1. One of these infected patients, before dying of AIDS, testified to the U.S. Congress that she had no risk factors for the disease (no drugs, no sexual activity, and no blood transfusions), leading her to believe that her infection must have come from Dr Acer during routine dental procedures. This soon led to a medical inquiry whereupon it was discovered, from molecular phylogenetic analyses, that the particular strain of HIV-1 possessed by the infected woman, as well as those of four of Dr Acer's other patients, were genealogically linked to HIV viruses isolated from Dr Acer (Fig. 6.11B). Indeed, the genetic resemblances between HIVs from all of these infected individuals were extremely close, like those normally observed only among HIV samples isolated from a single patient at successive time periods, or between the HIVs from a mother and her directly infected infant. These molecular genealogical findings provided the first genetic confirmation of HIV transmission (presumably unintentional) from an infected healthcare professional to clients.

# 7 | Geographical distributions

Geography is another "trait" that can be subjected to PCM. In this case, the geographical arrangements of species (i.e. their character states with respect to space) are plotted onto species' phylogenies as estimated from molecular or other genetic data. The usual intent is to reciprocally illuminate the geological histories of landforms (or bodies of water) and the evolutionary histories of organismal lineages that have inhabited those areas. For example, about three million years ago the Isthmus of Panama gradually emerged above sea level, creating a land bridge that facilitated movements of terrestrial organisms between North and South America and effectively blocking genetic interchanges between marine populations in the tropical Atlantic versus Pacific Oceans. The phylogenetic impacts of this geophysical event can be studied today by comparing molecular patterns among living species in that part of the world.

Strictly speaking, geographic features do not evolve (only biological entities do), but they certainly change through time in response to geological and other physical forces of the planet, and they certainly can leave major evolutionary genetic footprints on conspecific populations, closely related species, and broader taxonomic groups. Furthermore, the evolutionary pathways marked by these phylogenetic footprints often lead researchers to new discoveries about cryptic species or otherwise unrecognized biodiversity hotspots that can be highly important in conservation efforts.

## Afrotheria theory

For several decades, geologists have known that continents drift about the surface of the planet, sometimes moving apart and sometimes colliding with one another like bumper cars in a circular arena. The geophysical forces involved are so great that they periodically shake the earth and cause the gradual rise of great mountain chains such as the Himalayas, Rockies, and Andes. The movements of continental landmasses by plate tectonics are excruciatingly slow (typically about one inch (2.54 cm) per year as measured by precise instruments), but even at this snail's pace two continents drifting apart for 100 million years would open up a basin roughly 3200 miles wide, approximately the width of the North or South Atlantic Oceans.

Indeed, this is exactly what happened. For example, geological evidence indicates that South America and Africa were connected as recently as 105 million years ago. These huge landmasses were among the last-wedded remnants of Gondwanaland, a Southern Hemisphere supercontinent in the Mesozoic Era (the "Age of the Dinosaurs") that also included India, Australia, and Antarctica. As Africa and South America became physically separated by plate tectonic movements, so too did their terrestrial biotas. Thus, except for occasional instances of long-distance movements (e.g. via seed dispersal or via circuitous land routes through North America and Eurasia), most of the plants and animals on the African continent presumably have been evolving independently of those in South America for as much as the past 100 000 millennia.

One of the most dramatic phylogenetic footprints of this geographical disjunction was uncovered only recently. To nearly everyone's surprise, a wide array of African mammals conventionally thought to be unrelated to one another now appear to constitute an ancient monophyletic group that arose and diversified on the isolated African continent. These animals include elephants, aardvarks, and hyraxes (all previously thought to be close relatives of other hoofed mammals), the tiny elephant shrews (heretofore classified as cousins of rodents or rabbits), golden moles and tenrecs (formerly thought to be allied with other insectivorous mammals such as shrews and moles), and sea cows. Previously, almost no one had even imagined the possibility that such morphologically diverse African creatures might belong to a single clade, but that is precisely what recent molecular phylogenetic analyses have consistently indicated (Fig. 7.1). This clade, which encompasses approximately one-third of the 20 extant mammalian taxonomic orders, is now considered by many researchers to be a superorder: "Afrotheria."

The Afrotheria theory implies that the traditional classifications based on morphology were incorrect if construed as phylogenetic summaries. In other words, earlier scientists had mistakenly interpreted various morphological features (such as presence of hooves in aardvarks, or the mouse-like body forms of elephant shrews) as documenting shared ancestries with non-afrotherian groups (such as antelopes and horses (Ungulata) and mice (Rodentia), in these two cases). In the light of the new molecular evidence, such phenotypic features would now be deemed phylogenetically misleading in the sense that they probably evolved independently in afrotherian and various non-afrotherian lineages. The Afrotheria theory also implies that during perhaps the past 75–100 million years, a tremendous diversity of morphologies has evolved within one large clade of African mammals.

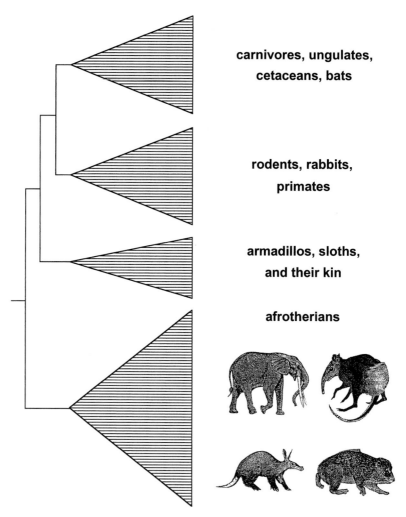

*Figure 7.1.* Broad-scale molecular phylogeny for placental mammals showing several suspected deep forks, including one leading to the superorder Afrotheria (after Eizirik *et al.*, 2001). The afrotherian animals pictured, clockwise from upper left, are an elephant, elephant shrew, aardvark, and hyrax. Drawings are reproduced with permission from Jonathan Kingdon.

Who would have guessed, for example, that massive elephants and tiny golden moles are phylogenetic cousins?

From similar kinds of molecular genetic evidence, the broader evolutionary tree of placental mammals displays at least three other deep branches (Fig. 7.1): one that includes hoofed animals (ungulates), whales and their allies (see *Cetacean origins*; Chapter 5), bats, and carnivorous groups such as cats and dogs; a second that includes rodents, rabbits, and primates; and a third composed of armadillos, sloths, and their kin. Molecular phylogeneticists now suspect that all of these groups split from one another and began their respective adaptive radiations during the later stages of the Mesozoic Era, some 65–150 million years ago. This time period roughly coincides with the breakup of Gondwanaland by the earth's crustal movements. By sundering and isolating Southern Hemisphere landmasses, including Africa, this continental breakup may thus have played a pivotal role in producing several early forks in the mammalian tree.

Not everyone accepts all aspects of the Afrotheria hypothesis, however. For example, Zack *et al.* (2005) used fossil and other evidence to suggest an American rather than African origin for a (somewhat redefined) afrotherian lineage. If so, not only might the current name for this clade be problematic, but so too might be the conclusion that the clade's genesis traces strictly to the breakup of Gondwanaland and the longstanding separation of the African continent from South America. Although such controversies surrounding the Afrotheria theory are not yet resolved to everyone's satisfaction, the fact that these issues are hotly debated is testimony to their inherent interest.

## Aussie songbirds

When European naturalists reached Australia a few centuries ago, they encountered many types of native bird that seemed rather familiar to them. There were fairy wrens: tiny avian sprites with trilled songs, saucy manners, and an oft-cocked tail, much like the wrens in Europe. Thornbill species in Australia were reminiscent, in behavior and appearance, of Old World warblers in English gardens. Australian sittellas spiraled down tree branches in search of bark insects, much as did tree-circling nuthatches back home. And Australian treecreepers, who probed the bark while spiraling up tree trunks, reminded the Europeans of the brown creepers that similarly work the trees in England.

So, the European naturalists understandably assigned these and other Australian birds to various taxonomic families where they seemed best to fit. For example, sittellas were put into the Northern Hemisphere nuthatch family (Sittidae), Australian treecreepers were placed in the Eurasian/American creeper family (Certhiidae), and fairy wrens were sometimes classified in the conventional family of wrens (Troglodytidae). Such was the traditional state of Australian and global avian systematics until the early 1980s.

Then along came ornithologists Charles Sibley and Jon Ahlquist. Using a new molecular technique known as DNA–DNA hybridization, these scientists began a phylogenetic reanalysis of numerous avian taxa from around the world, eventually

New Holland Honeyeater and Rufous Fantail

obtaining DNA samples from about 1700 of the world's 10 000 living bird species. As it turned out, the DNA–DNA hybridization data basically overturned conventional wisdom about several deep branches in the avian phylogenetic tree. In particular, the molecular findings indicated that many of Australia's native birds were not closely related to species elsewhere with similar appearances, behaviors, or lifestyles. Instead, the data suggested that Australian species as diverse as fairy wrens, thornbills, sittellas, and treecreepers (and many more, including scrub-birds, fantails, whistlers, woodswallows, pardalotes, and honeyeaters) were phylogenetically closer to one another than to their respective look-alikes on other continents (Fig. 7.2). In other words, many of Australia's songbirds seem to have evolved from a shared ancestor. This would mean that the Australian avifauna had diversified from common stock to occupy a wide variety of ecological niches on that continent. It would mean also that, as a by-product of this adaptive radiation, Australian and non-Australian birds had sometimes converged on one another morphologically and behaviorally, thereby confusing earlier taxonomists, who, by failing to appreciate this evolutionary phenomenon, erected improper classifications.

These molecular data also suggested that the evolutionary history of much of Australia's avifauna roughly paralleled that of its marsupial (pouched) mammals. Biologists have long been impressed by the evolutionary radiation of marsupials in Australia, and by the fact that various marsupial lineages had converged in appearance and lifestyle on placental mammals on other continents. For example,

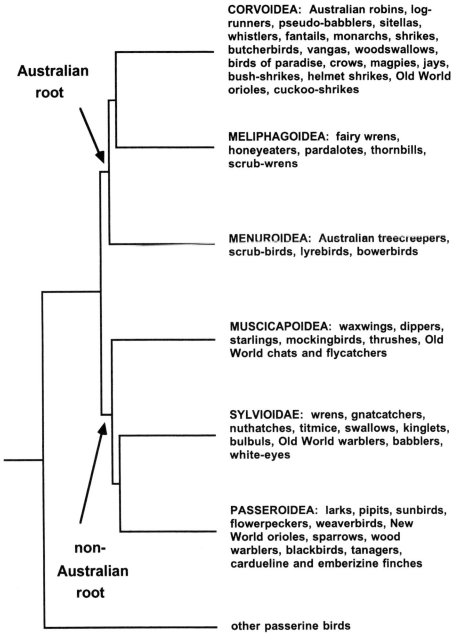

**Australian root**

CORVOIDEA: Australian robins, log-runners, pseudo-babblers, sitellas, whistlers, fantails, monarchs, shrikes, butcherbirds, vangas, woodswallows, birds of paradise, crows, magpies, jays, bush-shrikes, helmet shrikes, Old World orioles, cuckoo-shrikes

MELIPHAGOIDEA: fairy wrens, honeyeaters, pardalotes, thornbills, scrub-wrens

MENUROIDEA: Australian treecreepers, scrub-birds, lyrebirds, bowerbirds

MUSCICAPOIDEA: waxwings, dippers, starlings, mockingbirds, thrushes, Old World chats and flycatchers

SYLVIOIDAE: wrens, gnatcatchers, nuthatches, titmice, swallows, kinglets, bulbuls, Old World warblers, babblers, white-eyes

PASSEROIDEA: larks, pipits, sunbirds, flowerpeckers, weaverbirds, New World orioles, sparrows, wood warblers, blackbirds, tanagers, cardueline and emberizine finches

**non-Australian root**

other passerine birds

*Figure 7.2.* Phylogeny of oscine songbirds based on DNA hybridization data (after Sibley and Ahlquist, 1986). These researchers postulated the existence of two major historical groups: Corvida (of Australian ancestry, but sometimes with secondary evolutionary radiations in other parts of the world); and Passerida (with non-Australian evolutionary roots).

herbivorous kangaroos are more or less the ecological equivalents of placental deer, marsupial wombats are ecologically like placental woodchucks, and the marsupial Tasmanian wolf (now extinct) filled the carnivorous role of placental wolves in the Northern Hemisphere. However, thanks to a brood pouch (marsupium) and other distinctive traits shared by all marsupials, systematists always understood that a phylogenetic unity underlies the outward morphological diversity of these Australian mammals. It now appears that a comparable situation applies to many of Australia's songbirds.

The reason probably has to do with the fact that Australia has been one of the planet's most secluded landmasses over much of the past 100 million years, following the ancient breakup of Gondwanaland. In this isolated geographic setting, songbirds (like marsupial mammals) apparently flourished and diversified, often converging in form and behavior on unrelated creatures elsewhere in the world. The phylogenetic footprints of these evolutionary processes offer another compelling example of how geological forces such as plate tectonics and continental drift, as well as natural selection, have shaped the planet's biotas.

Recent studies based on more direct analyses of DNA sequences (see, for example, Barker *et al.*, 2004) have corroborated Sibley and Ahlquist's (1986) findings to some extent, but also significantly modified some of the original conclusions. Things now appear more complicated than originally supposed, in part because multiple waves of songbird dispersal from different centers of origin and at various evolutionary times have probably muddied the picture to some degree. None the less, the molecular data still point to the Australasian region as having been the site of a major evolutionary radiation of songbirds, the phylogenetic repercussions of which remain registered today in the genomes of living songbirds. What remains perhaps most remarkable is how profoundly the modern biological world has been fashioned by ancient geological events.

### Madagascar's chameleons

Chameleons (family Chamaeleonidae, subfamily Chamaeleoninae) are familiar reptiles with several peculiar and distinctive features including the following: a protractible tongue that is about as long as the animal's body; opposable toes in which the digits are fused in a way that allows the animal to grasp small branches as if by handshakes; a laterally flattened body; the frequent presence of horns or crests on the head; eyes mounted on protruding cones that can be moved independently of one another; a prehensile tail (in arboreal species); an ability to change body color quickly; and an odd locomotory behavior in which the creature rocks slowly back and forth after each step.

Madagascan chameleon

The global arrangements of chameleons have intrigued biologists almost as much as have their special morphologies and behaviors. Approximately 150 living species of chameleon inhabit various parts of Africa, the Indian subcontinent, Madagascar, and several archipelagos (such as the Seychelles and the Comoros Islands) in the Indian Ocean. There is a near unanimity of opinion that Chamaeleoninae is a monophyletic taxon, but just how these land-based species came to occupy multiple continents and islands is more controversial. One obvious hypothesis is that chameleons occasionally hitched oceanic rides on drifting rafts of floating vegetation. Via such chance over-water dispersal from ancestral homelands, chameleons at various times in the past presumably colonized new and open terrain where subsequent adaptive radiations sometimes took place.

An alternative hypothesis is that ancestral chameleons hitched rides on drifting continental landmasses as they floated slowly across the earth's surface over geological time. About 200 million years ago, all continents of the Southern Hemisphere were amalgamated into one grand landmass (Gondwanaland), which then slowly began to break apart by plate tectonic movements of the earth's crust. First came a basal split of South America plus Africa from the other southern protocontinents. Next, more than 150 million years ago, India–Madagascar began to separate and drift northward from another landmass composed of Antarctica, Australia, and New Zealand. In turn, both of these landmasses likewise fractured and gradually drifted apart, the modern continents eventually reaching their present configurations. Of special relevance to the current discussion was the fate of India–Madagascar. About 90 million years ago, India separated from Madagascar and began to drift northward, eventually colliding (about 60 mya) with Eurasia and thereby initiating the upward thrust of the Himalayan Mountains.

The oceanic-dispersal and continental-drift hypotheses make different predictions about the phylogenetic histories of chameleons. Under the former scenario, even the deepest nodes in the chameleon tree should postdate the geological rifts between landmasses driven by plate tectonics. Under the latter scenario, by contrast, some of the deepest nodes in the phylogenetic tree should be well more than 100 million years old, and, furthermore, the branching history of the major genetic lineages should faithfully mirror the geological history of the continental breakups. Thus, the oldest chameleon separation would be between African and Indian–Madagascan lineages, followed later by a separation of Madagascan lineages from those in India.

To test these competing hypotheses, Raxworthy *et al.* (2002) reconstructed a phylogenetic tree (based on molecular and other characters) for more than 50 species of chameleon from Africa, Madagascar, India, and islands of the Indian Ocean (Fig. 7.3). Their findings supported the oceanic-dispersal scenario, and refuted the continental-drift hypothesis, in at least two major regards. First, the chameleon phylogeny did not match the physical "area cladogram" of the landmasses involved. An area cladogram is a branching summary of the geophysical histories of landforms or bodies of water, and in this case it would depict a basal split of Africa from India–Madagascar, followed later by an historical separation between Madagascar and India. The chameleon phylogeny does not conform well to this geophysical arrangement. Instead, it shows a basal phylogenetic split of chameleon lineages within Madagascar, a phylogenetic intermingling of African and Madagascan lineages elsewhere in the tree, and a derived position of an Indian lineage *vis-à-vis* several of those in Africa (Fig. 7.3).

Second, the estimated dates in the chameleon phylogeny did not match predictions of the continental-drift model. Instead, based on molecular-clock calibrations for mtDNA sequences, even the deepest nodes in the chameleon tree were far more

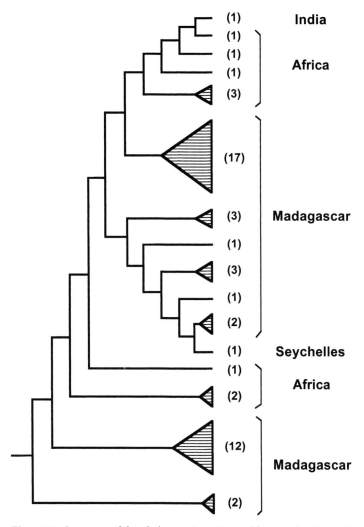

*Figure 7.3.* Summary of the phylogeny (as estimated from molecular and other data) and present-day geographic distributions of chameleons in the Indian Ocean region (after Raxworthy *et al.*, 2002). Numbers of genetically surveyed species in each clade are indicated in parentheses. Note that this phylogeny differs dramatically from the area cladogram (see text) of historical landmass relationships.

recent (*c.* 30–70 mya) than the geological separations of the relevant continental landmasses (> 90 mya). Other evidence likewise was more consistent with the oceanic-dispersal hypothesis. For example, the volcanic Comoros Archipelago is less than five million years old and it was never physically connected to mainland sources, so chameleons must have arrived there recently by over-water dispersal.

Thus, from the observed structure of the molecular phylogeny as well as other evidence, Raxworthy and colleagues concluded that chameleons probably underwent a post-Gondwanaland evolutionary radiation beginning in Madagascar, followed by out-of-Madagascar oceanic dispersal to Africa, to smaller islands of the Indian Ocean, and eventually to India (via Africa). There were secondary radiations and perhaps some back-colonizations as well, but the point is that all of this evolutionary action occurred long after the vicariant separations of the landmasses themselves.

Similar PCM analyses in a geographic context have been conducted on several other groups of Madagascan animals ranging from ants and spiders to frogs, snakes, rodents, and primates. Most of the findings also point to over-water dispersal as a primary contributor to biogeographic distributions in the region, although the phylogenetic details often differ. For example, by using logic similar to that described above for chameleons, Roos *et al.* (2004) concluded that strepsirrhine primates (lemurs and their relatives) originated in Africa and subsequently colonized Madagascar and Asia by respective single immigration events (probably by rafting). Relatively recent overseas dispersal events in the Indian Ocean region have also been deduced, by phylogenetic analyses, to be quite common in some amphibian groups (Vences *et al.*, 2003).

On the other hand, at least one recent report (Biju and Bossuyt, 2003) has identified a phylogenetic pattern consistent with a continental-drift model of ancient lineage separations. A newly discovered burrowing frog (*Nasikabatrachus sahyadrensis*; Nasikabatrachidae) from India proved upon phylogenetic analysis to be the sister taxon to another frog family (Sooglossidae) known only from the Seychelles Archipelago (formerly part of the India–Madagascar landmass). Furthermore, molecular sequence data (from nuclear and mitochondrial DNA) suggested that the phylogenetic split between Nasikabatrachidae and Sooglossidae occurred approximately 130 mya, roughly coincident with the relevant breakup of Gondwanaland.

## The evolutionary cradle of humanity

In addition to illuminating the phylogeographic origins of major species assemblages (as illustrated for the higher mammalian, avian, and reptilian taxa in the three preceding sections), PCM analyses can also help to identify the evolutionary birthplace(s) of individual species, one at a time. The approach involves estimating a molecular phylogeny for conspecific individuals or populations sampled throughout the current range of the species in question, and then interpreting results in concert with other lines of evolutionary evidence (such as the paleontological record) to reveal that species' ancestral homeland.

Chimpanzee and human

No species has received more attention in this regard than *Homo sapiens*. From compelling molecular and other evidence, chimpanzees are undoubtedly humans' closest living relatives, with the phylogenetic split between the two lineages occurring roughly five million years ago. This does not mean, however, that full-blown humans (or chimps) originated at that time, but rather that proto-human and proto-chimp lineages then separated from a common ancestor that was neither modern human nor modern chimp. Based on fossil evidence, more than four million years would pass in the hominid lineage before creatures morphologically indistinguishable from present-day humans eventually walked onto the evolutionary stage. Where on earth did that theatrical entrance take place?

Today, humans are globally distributed, but our species must have originated somewhere before migrating outward to people the planet. Previously, some (not all) paleontologists argued that pre-human lineages arose in several different regions of the world more than a million years ago, and remained completely isolated from one another until recent times. Molecular phylogenetic analyses have painted a very different picture, indicating instead that modern humans (*Homo sapiens sapiens*) originated once and only once, on the African continent, within the past few hundred thousand years. From that evolutionary cradle of origin, modern humans then spread eventually throughout the world, replacing (or perhaps interbreeding to some extent with) more archaic hominid populations already present elsewhere.

The first strong evidence for this recent "out of Africa" scenario came from studies of human mitochondrial (mt) DNA. The mitochondrial genome in most animals, humans included, is transmitted across the generations strictly through female lineages or matrilines, in contradistinction to most genes in the cell's nucleus, which are passed to progeny via parents of both sex. In an already classic paper, Rebecca Cann and colleagues (1987) surveyed mtDNA sequences from native peoples around the world in order to examine the current geographic distributions of various human matrilines. Three salient findings emerged from their phylogenetic analysis (Fig. 7.4). First, the root (basal node) of the matrilineal tree most likely existed in Africa, because only native Africans are represented on both major branches (A and B) of the current global mtDNA phylogeny. Second, much greater matrilineal diversity occurred in African populations than in native peoples from any other continental region. Third, based on molecular clock considerations, the entire mtDNA tree was temporally shallow, its deepest node dating back just a few hundred thousand years. The first two results point toward Africa as being the most probable site of origin for human matrilines that have survived to the present. The third result implies that the African matrilineal great-great-great . . . grandmother for all of current humanity lived a mere ten or twenty thousand generations ago.

Did these findings definitively settle all questions about human origins? No, especially because matrilines represent only a minuscule fraction of the total hereditary history of any species. So, researchers turned next to phylogenetic studies of Y-chromosome genes, which by virtue of their transmission strictly through males provide a record of patrilineal history. Results were generally similar to those for mtDNA, again pointing to a rather recent origin of modern humans (whether in Africa or elsewhere was less certain). Did this definitively settle the issue? No, especially because the matrilineal + patrilineal components of humans' extended pedigree still represent only a tiny fraction of our total hereditary history. Most of our genes are housed on autosomes (chromosomes in the cell nucleus other than the Y and the X) and are thus transmitted across multiple generations via both males and females. So researchers then turned their molecular phylogenetic analyses to a wide variety of autosomal genes. Most of these studies (with some

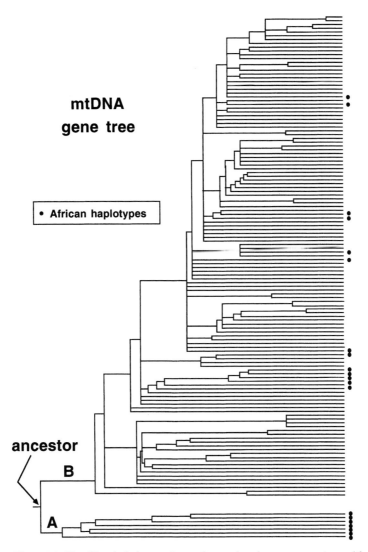

*Figure 7.4.* Matrilineal phylogenetic tree for modern humans as estimated from mtDNA sequence data (after Cann *et al.*, 1987, as redrawn by Avise, 2000). Black dots indicate genealogical positions of mtDNA genotypes observed in native Africans. Genotypes of native Asians, Australians, Europeans, and New Guineans are generally scattered throughout branch B of the mtDNA gene tree.

possible exceptions) again seemed to argue for a recent African origin for modern humans.

Based on these molecular findings from PCM, the popular press reported that all humans alive today had a matrilineal "Eve" and a patrilineal "Adam" who lived approximately 200 000 years ago in an African "Garden of Eden." This Biblical-like headline is correct to some extent, but it is also oversimplified and misleading if interpreted to imply that literally only two people lived back then. Mathematical calculations and biological reasoning suggest instead that the Garden of Eden probably was populated by at least several tens of thousands of individuals, many of whom made genetic contributions to modern humanity. This may seem counterintuitive at first, but it is a logical consequence of the fact that our matrilineal Eve and patrilineal Adam were at the root of only two of multitudinous genetic pathways in our collective ancestry. One way to see this is by considering your own genetic heritage going back (for example) across just three generations. You received your mtDNA from your matrilineal great-grandmother, and your Y-chromosome (if you are a male) from your patrilineal great-grandfather. But you actually had eight great-grandparents, all of whom contributed in more or less equal proportions to your total nuclear genetic ancestry.

### Coral conservation

Sibling species are closely related forms of life that appear nearly identical in morphology yet whose populations are reproductively isolated from one another (as judged by their genetic distinctions, for example, or perhaps their co-occurrence in sympatry). The sea may be chock full of previously unrecognized sibling species. Such was the dramatic conclusion reached in 1993 by Nancy Knowlton, a marine biologist who for years had been conducting population genetic surveys on a wide variety of invertebrate taxa. From her molecular studies, and others like them, evidence had accumulated that, in taxonomic assemblages ranging from sponges and corals to marine worms, mollusks, urchins, barnacles, and many others, cryptic species abounded. Such findings are sometimes bolstered by more detailed morphological or behavioral appraisals. To pick one example, close scrutiny of the behaviors and other features of copepods (tiny crustaceans) in the genus *Tisbe* raised the number of recognized species from just a few to more than 60 (Marcotte, 1984). The discovery that sibling species are rife among marine invertebrates clearly has important ramifications for basic ecological and evolutionary studies as well as for more applied fields such as conservation biology.

About a decade later, Knowlton and her colleagues (Fukami *et al.*, 2004) were at it again, this time reporting that not only cryptic species, but also cryptic phylogenetic lineages and clades, may be common in the sea. A primary example involves corals. Based on morphological considerations, more than 100 genera of reef-building coral had been recognized, most consisting of multiple species and each presumably being of monophyletic origin. That latter presumption may be incorrect in some cases, according to Knowlton and her crew based on their

*Montastraea* coral

molecular phylogenetic findings. Using DNA sequences from nuclear and mito-chondrial genes, they estimated a phylogeny for more than 80 coral species rep-resenting about 30 genera, and then plotted both the traditional generic assign-ments and the species' geographic ranges onto that phylogenetic tree (Fukami *et al.*, 2004). To their surprise and that of the marine scientific community, con-generic species often failed to form clades, and indeed sometimes were scattered widely across the phylogeny (Fig. 7.5). Conversely, some species previously placed in separate genera and thought not to be close kin occasionally joined one another in clades that today are confined to a specific oceanic basin (Atlantic or Indo-west Pacific).

For example, in terms of their DNA sequences, species of *Favia* and *Scolymia* in the Atlantic region proved to be phylogenetically closer to one another than to their respective congeners in the Indo-Pacific (Fig. 7.5). Similarly, upon molecular phylogenetic analysis the genus *Montastraea* now appeared to be an artificial (i.e. polyphyletic) collection of unrelated species masquerading under a similar external

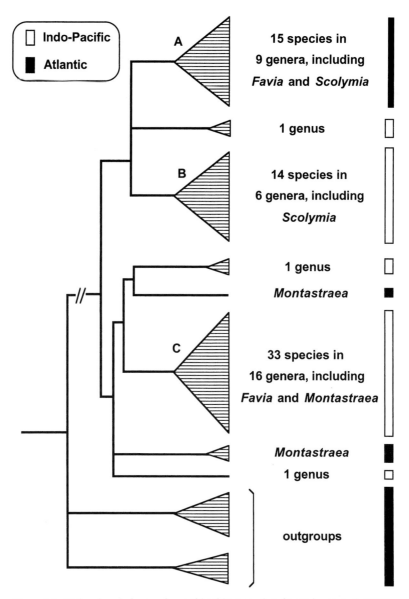

*Figure 7.5.* Molecular phylogeny for reef-building corals (after Fukami *et al.*, 2004). Note how the species and genera in clades A, B, and C group according to oceanic basin rather than to their traditional morphology-based taxonomic assignments.

phenotype. From these and other such examples, Knowlton's group concluded that clade membership in reef-building corals was better predicted by geography than by gross morphology. Apparently, pervasive evolutionary convergences in coral external morphology had confused earlier systematists and led to them to erect classifications that were not accurate reflections of coral phylogeny.

These molecular phylogenetic discoveries have ramifications for conservation as well as taxonomy. A common sentiment is that highly distinctive evolutionary lineages warrant special preservation efforts because they contribute dispropor-tionately to life's overall genetic diversity (Mace *et al.*, 2003). In the current case, a distinctive clade of Atlantic corals was identified (Fig. 7.5) that significantly alters our perceptions about global coral diversity. Under the traditional (and apparently misguided) coral classification scheme, only 17% of reef-building genera and no coral families were considered endemic to the Atlantic basin, whereas the cor-responding figures for the Indo-west Pacific were 76% and 39%. In other words, the Atlantic region was thought to be a relatively unimportant evolutionary cra-dle for distinctive coral lines. However, based on the molecular evidence, Fukami *et al.* (2004) estimate that the formerly cryptic lineage of Atlantic reef-building corals separated from Indo-Pacific lineages about 35 million years ago to become the monophyletic wellspring of an impressive adaptive radiation in the Caribbean region. During that diversification process, various Caribbean sublineages appar-ently converged on various Indo-west Pacific corals in terms of general external morphologies and lifestyles. If these new evolutionary interpretations are correct, it means that Caribbean corals should be cherished even more when setting con-servation priorities.

## Sri Lanka, a cryptic biodiversity hotspot

Sri Lanka (formerly known as Ceylon) is a large island in the Indian Ocean, just off the southeastern tip of the Indian subcontinent. Over the past two million years, however, it has been connected to the mainland repeatedly. During each of several Pleistocene Ice Ages, vast quantities of the earth's surface waters lay frozen in continental glaciers, causing sea levels to drop worldwide by more than 100 m and (among many other effects) exposing a broad isthmus or land bridge that joined Sri Lanka with India. Thus, the current separation of the island from the mainland dates only to about 10 000 years ago, when the most recent Ice Age glaciers melted and global sea levels rose once again to create the Palk Strait (the present-day shallow marine channel between Sri Lanka and the Indian subcontinent).

Traditionally, biogeographers thought that Sri Lanka's terrestrial and freshwa-ter biotas were closely related to those in adjoining areas of mainland India. This presumption stemmed not only from recent physical connections between the two bodies of land (which should have allowed many opportunities for interregional dispersal and gene flow), but also from the fact that representatives of several taxonomic groups of Sri Lankan and Indian animal showed high levels of morphological similarity suggestive of very recent (e.g. late-Pleistocene) shared ancestry. In many cases, these morphological resemblances were so close that systematists had classified particular Sri Lankan and Indian forms (such as populations of various frogs, and of some snakes) as conspecifics. A close affiliation between the two biotas was also reflected in the fact that Sri Lankan faunas and those of the Western Ghats region of southern India were deemed to compose a single biodiversity hotspot (i.e. a rich community of species that fits together as a biogeographic unit). This was considered one of the top 25 biogeographic hotspots around the world that have been deemed to be especially fragile and worthy of protection (Myers *et al.*, 2000).

Recent phylogenetic discoveries from DNA-level appraisals have extended some aspects of these conventional notions. By analyzing mtDNA sequences gathered from dozens of species of frogs, snakes, freshwater crabs and fishes, and other taxonomic groups, Bossuyt and colleagues (2004) uncovered previously unsuspected phylogenetic distinctions between Sri Lankan and Indian taxa. For example, among caecilians (legless amphibians that look superficially like snakes), five *Ichthyophis* species from Sri Lanka proved to belong to a single monophyletic group well distinguished from related clades in southern India or Southeast Asia (Fig. 7.6). Similarly, among tree frogs primarily of the genus *Philautus*, about 20 surveyed species from Sri Lanka proved to belong to a recognizable clade that included only a few of the 16 surveyed species native to southern India (Fig. 7.6). Generally similar phylogenetic patterns emerged for shieldtail snakes in the family Uropeltidae, fishes in Cyprinidae (genus *Puntius*), crabs in Parathelphusidae and Gecarcinucidae, and freshwater shrimps in Atyidae (genus *Caridina*). In each case, most if not all species native to Sri Lanka belonged to a monophyletic group that was historically allied to, but none the less recognizably distinct from, counterpart species and clades found in southern India.

These results provided nearly incontrovertible evidence for local and previously cryptic endemism within the broader Sri Lanka – Western Ghats biodiversity hotspot. In particular, Sri Lanka appears to have been the site of extensive adaptive radiations *in situ* in several taxonomic groups. Although a few evolutionary lineages appear to have recently colonized southern India from Sri Lanka (see right-hand side of Fig. 7.6), and vice versa, many of Sri Lanka's clades show a high degree of insularity. These phylogeographic findings identify Sri Lanka and the Western Ghats as biodiversity subregions that are distinct enough from one another to warrant special management efforts to preserve their rich and historically unique biotas.

## caecilians                                    frogs

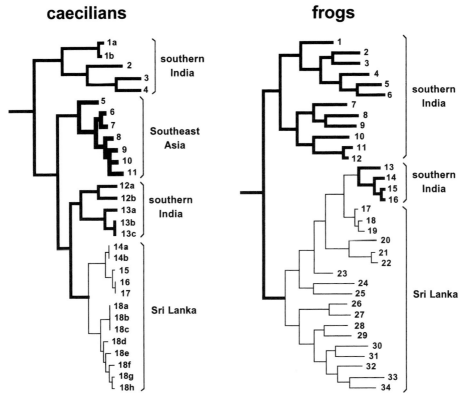

*Figure 7.6.* Molecular phylogenies for caecilian amphibians in the genera *Ichthyophis* and *Uraeotyphlus* (left), and tree frogs in the genus *Philautus* (right), in the regions of southern India and Sri Lanka (after Bossuyt *et al.*, 2004). Specimens representing different species (some not yet formally described) and different populations are indicated by numbers and letters, respectively. Heavy, medium, and thin lines in the trees signify lineages in Southeast Asia, southern India, and Sri Lanka, respectively.

A question remains, however, as to why the Pleistocene land bridges between Sri Lanka and southern India did not provide ready thoroughfares for extensive faunal interchanges. If species exchanges had taken place frequently, any phylogenetic distinctiveness of the Sri Lankan and Indian faunas should have been blurred and few if any region-specific clades would be evident today. One plausible explanation is as follows. The taxa genetically surveyed by Bossuyt *et al.* (2004) are inhabitants primarily of wet zones in Sri Lanka and moist forests in the Western Ghats. However, situated between these rainforest environments are extensive arid lowlands that probably served during the Pleistocene (as well as now) as powerful dispersal barriers for rainforest-adapted species.

### Overseas plant dispersal

The plant family Goodeniaceae (order Asterales) contains about 400 species of tropical herbaceous shrub and small tree. Most of the 11 recognized genera within the family are confined almost entirely to Australia, where they constitute significant elements of that continent's coastal flora. However, one genus – *Scaevola*, with 130 species – has dispersed and radiated throughout much of the Pacific region, and a few of its species have even found their way into tropical Africa, Madagascar, Sri Lanka, southern India, and the Americas including the Galápagos Islands.

Several interesting questions surround this global pattern of evolutionary diversification from an original Australian source. What traits are possessed by particular species that enabled them to traverse vast expanses of ocean and colonize distant continents and isolated islands (such as the Hawaiian Archipelago)? How many successful transoceanic dispersal events altogether were involved? And were some of the primary colonizations followed elsewhere by secondary adaptive radiations? In particular, two competing hypotheses have been advanced to explain the extensive proliferations of *Scaevola* species outside of Australia: (a) each island endemic and continental species arose from a separate out-of-Australia colonization event; or (b) only one or a few such colonizations were involved, but each was followed by speciation episodes in the newly inhabited regions.

These and related issues have been addressed and provisionally resolved by using historical footprints in these plants' nuclear ribosomal DNA sequences. A molecular phylogenetic analysis by Howarth *et al.* (2003), involving 50 representative species (Fig. 7.7), led these researchers to the following conclusions. First, *Scaevola* species outside of Australia arose from at least six or seven primary colonization events (the exact number is uncertain because of less than perfect resolution in the phylogenetic tree). Four of these independent dispersal events each eventuated in a single extra-Australian species: *S. oppositifolia* in the Philippines (probably arriving via New Guinea), *S. gracilis* in New Zealand and Tonga, *S. beckii* in New Caledonia, and *S. glabra* in the Hawaiian Islands. The other three dispersal events appear to have been followed by secondary speciations giving rise to larger groups. These latter radiations led to the following (Fig. 7.7): various *Scaevola* endemics (all closely related to *S. taccada*, a widespread Pacific species) on South Pacific islands; several other *Scaevola* species in the Northern Hemisphere (all closely related to *S. plumieri*, a widespread Atlantic species); and the sister species *S. micrantha* and *S. chanii* on Borneo.

A second main conclusion was that the Hawaiian Islands had been colonized successfully on at least three separate occasions, each involving a different *Scaevola* lineage. The *S. glabra* lineage (colonization number 5 in Fig. 7.7) probably arrived

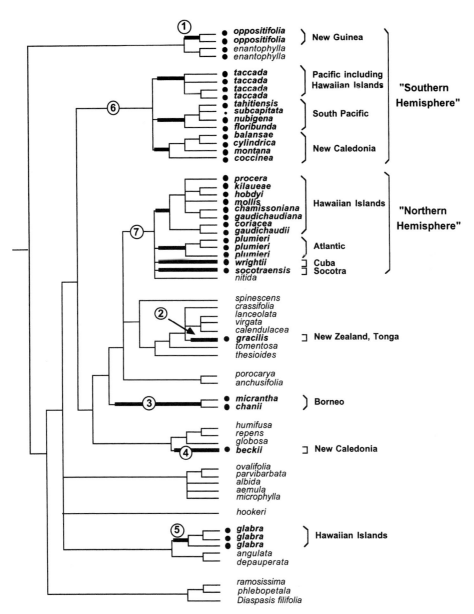

*Figure 7.7.* Molecular phylogeny and geographic distributions for 50 species of *Scaevola* (after Howarth *et al.*, 2003). Taxa in boldface occur outside Australia; others are Australian. A black dot indicates that a species has fully fleshy fruit. Bold lines and numbers indicate hypothesized dispersal events from Australia to outside locations (see text).

in Hawaii directly from Australia, as judged by the geographic distributions of *S. glabra*'s closest living relatives. The *S. taccada* lineage (colonization event 6) probably emigrated from Polynesia, and the *S. plumieri* lineage (event 7) may have embarked from the American Pacific coast. This *S. plumieri* colonization event also appears to have been followed by an evolutionary radiation *in situ*, leading to several Hawaiian endemics (Fig. 7.7). Thus, overall, the current phylogenetic diversity of *Scaevola* species in the Hawaiian Archipelago reflects multiple immigration events as well as several local speciations. This contrasts with the situation in most other species-rich taxonomic groups endemic to these islands. In Hawaiian silverswords and lobelioid plants, for example, as well as in Hawaiian honeycreeper birds and fruitflies, similar molecular phylogenetic analyses (see References) have indicated that each adaptive radiation occurred within the Archipelago itself, beginning with a single founding lineage.

A third issue phylogenetically addressed by Howarth *et al.* (2003) concerns the mode of overseas dispersal in *Scaevola*. An important clue came from the unexpected distribution of fleshy (drupaceous) fruits, as opposed to dry non-fleshy fruits, on the molecular phylogeny (Fig. 7.7). From traditional classifications (based in part on how the plants' seeds are housed), the fleshy-fruit condition, characteristic of about one-half of the *Scaevola* species surveyed, was thought to have arisen perhaps only once within the genus. However, the new estimate of *Scaevola* phylogeny, based on molecular data, implies that the fleshy-fruit state arose on multiple independent occasions. Furthermore, as shown in Fig. 7.7, all surveyed species of *Scaevola* now residing outside Australia belong to lineages characterized by fleshy fruits.

This association between overseas distributions and the nature of the fruit probably reflects an underlying causal relationship. Seeds encased in fleshy *Scaevola* fruits can remain viable while floating in seawater even for several months, and they are also known to survive passage through an avian gut. Either of these factors could predispose such seeds for long-distance dispersal. Indeed, Howarth *et al.* (2003) suggest that fleshy fruits were key evolutionary innovations that enabled particular *Scaevola* lineages to colonize foreign lands via transoceanic currents or migrating birds.

## Phylogenetic bearings on Polar Bears

When it comes to phylogenetic relationships, external appearances can be deceiving. Even with regard to some of the world's most familiar and charismatic animals, closer examinations of internal molecular genetic characters occasionally yield big phylogenetic surprises. The Polar Bear (*Ursus maritimus*) offers a fine example.

Polar Bear

With its thick snow-white coat and aggressive predatory nature, the carnivorous Polar Bear is well adapted for hunting seals and walruses in its frigid, vegetation-free, circum-Arctic homeland. Its omnivorous cousin the Brown Bear (*U. arctos*), known in some places as the Grizzly, is equally at home farther south, where its deep brown fur generally blends in with the forests and grasslands it roams. No one has doubted that Polar Bears and Brown Bears are genetic cousins at some level, but until recently no one had imagined just how close that phylogenetic relationship might be.

The Brown Bear's historic range encompassed most of the northern Holarctic Realm (North America plus northern Eurasia), where it still survives mostly in isolated regions left undisturbed by humans. Several independent research groups have collectively surveyed mtDNA sequences from hundreds of animals representing populations across this vast distribution. The molecular data showed that *U. arctos* is geographically divided into at least five highly distinct matriarchal clades, each currently confined to one of the following areas: western Europe, southern Canada plus the contiguous United States, northern Canada plus eastern Alaska, western Alaska plus Siberia plus eastern Europe, and the ABC Islands (Admiralty, Baranof, and Chichagof) of southeastern Alaska (Fig. 7.8). The spatial arrangements of these distinctive matrilines seem to make considerable sense when interpreted as reflecting how this species probably was sundered into separate glacial refugia at various times during the Pleistocene Epoch.

The big surprise came when *U. maritimus* was added to the analysis. Rather than being a sister taxon or outlier, Polar Bears were embedded well within the collective Brown Bear clade. In other words, with respect to matrilineal ancestry, Brown Bears appeared to be paraphyletic with respect to Polar Bears. More specifically, Polar Bears were extremely close kin to Brown Bears inhabiting the ABC Islands. The magnitude of the mtDNA genetic distance suggests that the Polar Bear lineage separated from ancestral Brown Bear lineages as recently as 200 000 years ago.

This molecular phylogenetic finding was completely unexpected; its interpretation still is under debate. One possibility is that secondary hybridization between the two species resulted in the transfer of mtDNA lineages from ABC Brown Bears to Polar Bears. This hypothesis might not be as implausible as it at first sounds, because in captivity the two species are capable of producing viable and fertile hybrid offspring. Another category of possibilities is that Polar Bears originated recently in evolution from ABC Brown Bear stock, or, conversely, that ABC Brown Bears arose very recently from ancestral Polar Bear stock. In either of these latter cases, adaptive evolution in one population or the other must have been rapid and extensive, involving for example not only a switch from brown to white fur (or vice versa), but also changes in relative neck lengths, head sizes, diets, behaviors, and general ecologies.

A third possibility is simply that the mtDNA genealogy might be misrepresentative of overall genetic relationship between Brown Bears and Polar Bears. More that 99% of the hereditary history of these (or any other) species lies in the vastly larger nuclear genome, which must now be mined for additional information before any definitive phylogenetic conclusions are drawn. Thus, further research will be required to decide whether the bears' external phenotypes or their currently available mtDNA sequences have produced the current phylogenetic enigma.

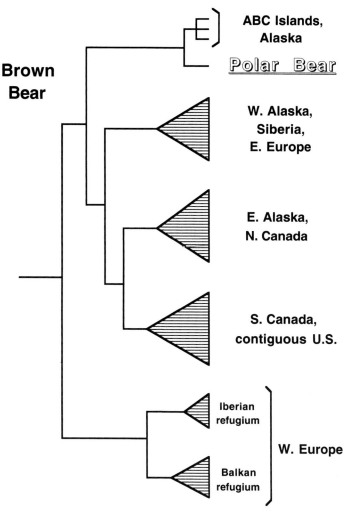

*Figure 7.8.* Global matrilineal phylogeography of Brown Bears and Polar Bears (after Avise, 2005, based on molecular studies by several researchers listed in References). Note the paraphyletic relationship of Brown Bears to Polar Bears in this mtDNA genealogy.

## Looking over overlooked elephants

Especially when a genealogy is reconstructed for spatially arrayed populations of the same or closely related species (as in the bear studies described in the previous section), the endeavor is termed a "phylogeographic" analysis. Literally hundreds of animal and plant species have been the subjects of molecular phylogeographic appraisals (Avise, 2000), but nowhere did a greater surprise emerge than in the apparent discovery of a new elephant species in Africa.

Traditionally, the Asian Elephant (*Elephas maximus*) and the African Elephant (*Loxodonta africana*) are placed in separate genera, but until recently taxonomists made no formal species-level distinctions among geographic populations on either of the two continents. The African Elephant occurs throughout much of sub-Saharan Africa, from tropical rainforests in the west-center of the continent to vast areas of open scrub and grasslands mostly to the east and south of the heavily forested region. Within that broad distribution, "forest elephants" were sometimes distinguished informally from "savannah elephants" by virtue of their habitat preferences, smaller size, longer and straighter tusks, smaller and more rounded ears, and flatter foreheads. However, not much was made of these ecological and morphological differences until recently when, with the benefit of molecular phylogenetic hindsight, they were reinterpreted by some authors to warrant recognition of two separate African species (*L. africana*, the savannah form; and *L. cyclotis*, the forest form).

In molecular assays of both mitochondrial and nuclear genes, initial genetic surveys found rather dramatic and unexpected differences between the forest and savannah elephants (Fig. 7.9). Based on the magnitude of the genetic differences, Roca *et al.* (2001) estimated that *L. africana* and *L. cyclotis* have been evolving independently for about 2600 millennia (give or take about a million years), and furthermore that the savannah elephant had probably undergone a population bottleneck (as judged by its diminished genetic variability) in rather recent evolutionary times. The forest and savannah elephants in Africa were not as divergent from one another as either proved to be from the Asian species, but none the less their genetic differences were both impressive and unanticipated.

A subsequent molecular study of additional geographic populations corroborated the general notion that distinctive elephant lineages exist in Africa today, but it also raised more surprises and some potential complications for the otherwise simple story described above. In particular, a third, highly divergent genealogical line was provisionally identified that included elephants inhabiting forests *and* savannahs of western Africa. The evolutionary etiology of this lineage (i.e. exactly how it arose and where it fits into the broader scheme of elephant relationships in Africa) is uncertain, pending further molecular genetic investigations. So too are the precise origins and historical demographies of several additional elephant sublineages in other regions of the African continent.

Whatever the final outcome may be, the emerging phylogeographic findings on African elephants are important for several reasons. First, they demonstrate that even some of the world's most conspicuous species can harbor substantial but formerly cryptic genealogical subdivisions. Second, they evidence the fact that, in terms of moving lineages from one geographic region to another over ecological

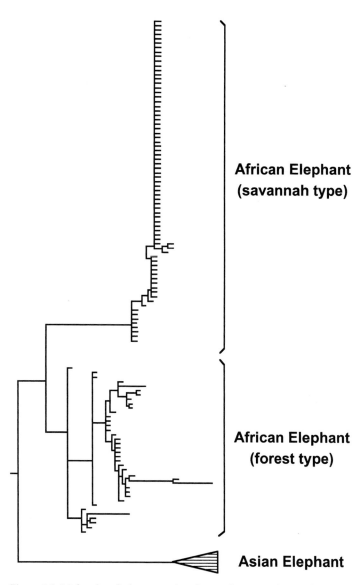

*Figure 7.9.* Molecular phylogeography of more than 100 African elephants (with Asian elephants as an outgroup) based on DNA sequences from four nuclear genes (after Roca *et al.*, 2001). Similar phylogeographic patterns (albeit with some notable differences; see text) have been reported in mtDNA sequences and microsatellite genetic markers (see Eggert *et al.*, 2002).

and evolutionary timescales, even highly mobile creatures such as elephants have sometimes fallen far short of their inherent dispersal potential. Third, they illustrate the point that taxonomy near the species level can be a challenging but also a somewhat subjective endeavor, especially when the genetically differentiated populations involved are allopatric (because whether or not they could hybridize successfully in nature often remains uncertain). Last but not least, they show the relevance of molecular phylogeographic patterns to conservation efforts. In this case, forest elephants in particular are severely threatened by poaching and habitat destruction, so their evolutionary distinctiveness should make the protection of these animals an even higher priority.

## Bergmann's rule

Ecogeographic rules are generalizations describing empirical relations between phenotypic traits and environmental variables. For example, Gloger's rule describes an observed tendency for species to be more darkly pigmented in regions with high humidity; Allen's rule notes a general trend wherein homeotherms (warm-blooded animals, also known as endotherms) often have relatively shorter appendages in geographical regions with colder climates; and the clutch-size rule notes a tendency for most birds to have larger clutch sizes at higher latitudes. In general, the possible ecological and evolutionary reasons for such inclinations are far more controversial than the empirical tendencies themselves, although the latter too are sometimes debated because many exceptions do exist.

Perhaps the best-documented ecogeographic rule – Bergmann's rule – was traditionally also thought to be the best understood in terms of adaptive significance. Bergmann's (1847) rule states that, within particular species of birds and mammals (the two major groups of endothermic animals), body size tends to be inversely related to ambient climatic temperature. The conventional explanation is that larger animals must be at a relative selective advantage in colder climates because their bodies, with lower surface-to-volume ratios than smaller-sized animals (all else being equal), inherently tend to conserve metabolically generated heat more efficiently. However, several recent surveys have cast doubt upon this interpretation by showing that some poikilotherms (cold-blooded animals, also known as ectotherms) also obey Bergmann's rule. In particular, most species of turtle (Ashton and Feldman, 2003) and salamander (Ashton, 2002) show patterns of geographic variation in which larger body sizes characterize populations in higher latitudes. On the other hand, most squamate reptiles (lizards and snakes) seem to obey the exact converse of Bergmann's rule: they often tend to have significantly smaller bodies in cooler climates (Ashton and Feldman, 2003).

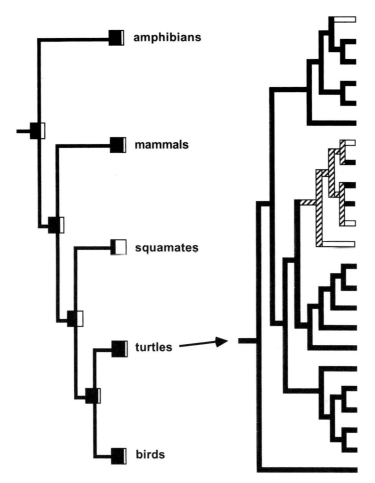

*Figure 7.10.* PCM analyses of Bergmann's rule in tetrapod vertebrates (after de Queiroz and Ashton, 2004). Left, probable ancestral states for major tetrapod groups. Relative volumes of black versus white in the squares indicate relative probabilities (as gauged by maximum likelihood reconstructions) that a given ancestral state was, respectively, the presence of Bergmann's rule or its converse. Right, probable ancestral states, as gauged by maximum parsimony reconstructions, for 23 surveyed species of freshwater and terrestrial turtle. Black branches indicate compliance with Bergmann's rule, white branches indicate its converse, and hatched branches indicate equivocal conditions.

Ecogeographic rules in general, and Bergmann's rule in particular, by definition describe collective properties of species rather than phenotypic features of individual organisms. In other words, a species can obey Bergmann's rule or its converse (or neither), but an individual animal by itself cannot. This point becomes important in the context of PCM analyses, which have now been conducted on several major groups of tetrapods (four-limbed vertebrates). In these PCM exercises, each species is first characterized as to whether it obeys Bergmann's rule or its converse, or neither. Then, by plotting these qualitative scorings on an independently generated phylogeny for the taxa in question, ancestral states of the ecogeographic rule are deduced or reconstructed on the tree. This approach has been applied to amphibians, turtles, squamates, birds, and mammals, as well as to all of these tetrapod groups considered collectively (de Queiroz and Ashton, 2004). Examples of these PCM reconstructions are provided in Fig. 7.10.

One evolutionary conclusion from such analyses is that the converse of Bergmann's rule was almost certainly the ancestral condition for squamate reptiles, whereas the presence of Bergmann's rule was the likely ancestral state for each of the following: turtles, amphibians, birds, and mammals. A second point of interest follows from the first, namely, that these intraspecific ecogeographic trends appear to have a cross-species evolutionary "heritability," meaning that whether or not a given species or clade obeys the rule is predicated to a large degree on whether or not its ancestors did so as well. A third point is that Bergmann's rule among the tetrapods apparently has an ancient origin that far pre-dates the evolutionary appearance of homeothermy.

This latter conclusion, in particular, strongly suggests that the traditional explanation for Bergmann's rule – that larger body sizes are adaptive for conserving internally generated body heat in cold climates – is inadequate to account for the initial evolutionary genesis of this broader ecogeographic trend among the vertebrates. Various possibilities suggest themselves. Perhaps Bergmann's rule actually has little or nothing to do with heat conservation, and an entirely different kind of biological explanation should be sought. For example, an ability to survive long periods with little food (rather than to conserve body heat *per se*) might conceivably account for why many poikilotherms as well as homeotherms have larger bodies in climatic regimes with higher seasonality and extended episodes of cold. Or perhaps the conventional thermoregulatory hypothesis is correct for homeotherms but another scenario applies to poikilothermic vertebrates, in which case Bergmann's rule may have been long maintained in evolution even as its adaptive significance shifted. Other examples of this general sort are known. For example, the nature of selection pressures changed dramatically as some of the jaw bones of ancestral tetrapods gradually became transformed into the ear bones of mammals.

As always, trait correlations and patterns as revealed by PCM cannot by themselves definitively diagnose causal evolutionary processes. They can, however, help to identify problematic evolutionary situations where new causal hypotheses might be developed and fruitfully tested.

# EPILOG

Before closing, I want to reiterate two disclaimers. First, although I have empha-sized the utility of *molecular* phylogenies as historical backdrops for interpreting organismal ecology and evolution, this was done primarily to provide a coher-ent theme and organizational framework for this book. In truth, phylogenies can also be successfully estimated by using all sorts of morphological, behavioral, and other organismal traits. Indeed, all phylogenies erected before the 1960s, and many since then, have been based on directly observable phenotypic traits rather than on proteins and nucleic acids. Usually, well-supported molecular phylogenies tend to agree with seemingly well-supported morphological phylogenies, as expected. Occasionally, however, they appear at face value to disagree; as I have tried to illus-trate by examples, resolution of the discord can be mutually illuminating about both molecular and organismal evolution. I have emphasized molecular phylogenetic approaches because they have offered exciting new perspectives on the biologi-cal world, and if I disproportionately spotlighted apparent phylogenetic conflicts between different types of data, it is only because these are the most scientifically interesting.

Second, for any or all of the case studies examined, the specific biological con-clusions reached (either by the original authors or by myself) remain provisional for several reasons. For example, there is ongoing debate about the relative phy-logenetic merits of different types of molecular data and their statistical analy-ses, controversies continue about precise historical relationships within many if not most of the taxonomic groups considered, and reservations typically abound about numerous details of the comparative phylogenetic approach and PCM analy-ses themselves (see Chapter 1 and the Appendix). The field of comparative phy-logenetics is now in an explosive growth phase as it enters the "era of genomics" with increasingly powerful laboratory and analytical methods for gathering and interpreting molecular data. Thus, I would not be surprised if some of the current biological inferences on specific taxa are called into question with new discoveries. Nor would I be unduly dismayed by these developments: such is the nature of sci-ence, especially in active frontier disciplines such as comparative phylogenetics.

Although particular inferences from current PCM studies may require modification or even abandonment as new and better information becomes available, PCM itself will remain a powerful interpretive tool for many types of ecological and evolutionary study.

Accordingly, my intent has been not so much to trumpet specific biological inferences from PCM analyses, but rather to illustrate how comparative phylogenetic perspectives can contribute to the process of biological discovery. In this important sense, the greatest joys in scientific research, as in other facets of life, often reside more in the quest than in the final resolution. Despite current blemishes and potential pitfalls, comparative phylogenetics offers a powerful new mode of inquiry into the evolutionary nature of nature. If this sentiment has been conveyed, and a greater interest has thereby been stimulated in the marvelous workings of the natural world, then this book will have served its purpose.

# A primer on phylogenetic character mapping

Many recent textbooks (see References for Chapter 1) thoroughly cover the laboratory techniques of molecular genetics as well as phylogenetic methods of data analysis, at levels suitable, depending on the book, for readers ranging from novice to
expert. Thus, this Appendix will confine its attention to some underlying concepts
and methods specific to PCM *per se*. In other words, it is assumed for current purposes that appropriate molecular genetic data have been gathered and properly
analyzed to estimate a robust phylogenetic tree for the taxa in question, and that
the intent now is to map and interpret the distributions of particular phenotypes
onto that tree. Normally, alternative states of phenotypic characters are known only
for extant species (exterior nodes) of the phylogenetic tree; the goals of PCM are to
infer ancestral character states at various interior nodes and to estimate character-
state transitions along various tree branches. Only elementary aspects of PCM will
be covered here. For far more complete and advanced treatments, including operational details, see Brooks and McLennan (1991, 2002), Eggleton and Vane-Wright
(1994), Harvey and Pagel (1991), Harvey *et al.* (1996), Maddison and Maddison
(2000), Page and Holmes (1998), and other references cited below.

## History of cladistic concepts and terminology

Systematists from Aristotle to Linnaeus and beyond traditionally grouped organisms and erected biological classifications based on qualitative or quantitative
appraisals of the overall resemblance (phenetic similarity) among taxa. Thus, in
effect, they failed to discriminate between two potential evolutionary sources of
organismal resemblance: patristic similarity (the component of phenetic resemblance due to shared ancestry), and homoplasy (the component of phenetic
resemblance due to convergence from separate ancestors). The net result in each
case was a biological classification that reflected some unspecified mix of genealogical history (phylogenetic signal) and homoplasy (phylogenetic noise).

In 1966, a book appeared that was to transform both the theory and practice of
systematics. It was an English translation of a work that had been written in 1950 by

the German entomologist Willi Hennig. In *Phylogenetic Systematics*, Hennig (1966) forcefully advanced the thesis that biological classifications should reflect phylogenetic relationships only. Thus began the Hennigian cladistic revolution that sought to make explicit distinctions between shared ancestry and convergence as evolutionary sources of organismal similarity, and thereby rectify a perceived major shortcoming of traditional systematics. Hennig developed his argument much farther, however. First, he developed the key notion that patristic similarity itself has two components: one attributable to shared derived traits (synapomorphies), and the other due to shared ancestral traits (symplesiomorphies). (These and other phylogenetic terms are defined in Box A1.) Second, Hennig explained why only genuine synapomorphic characters could properly identify monophyletic groups (clades), which he argued should alone be recognized in phylogeny-based classification schemes.

The basic logic (now universally accepted) underlying Hennig's latter two insights can be introduced by the following example involving three ingroup species (A–C) and one or more outgroup taxa (D). Neglecting D for the moment, and assuming that only bifurcations in a rooted phylogenetic tree are involved, the three ingroup species could be evolutionarily related in any of three ways (Fig. A1): ((A, B) C); ((A, C) B); or (A (B, C)). Suppose now that species A and B share a phenotypic character (e.g. wings) that C does not possess. This distribution of characters might be explained in any of several ways, depending on which cladogram (branching structure in a phylogenetic tree) is correct for the ingroup species, and which character state (winged or wingless) was the original ancestral condition for the ingroup (Fig. A1). For example, if A and C are sister taxa composing a clade to the exclusion of B (panels *c* and *d* in Fig. A1), then the shared presence of wings in A and B is attributable either to evolutionary convergence (e.g. panel *c*) or to the retention of an ancestral character state (panel *d*), depending, respectively, on whether wings are derived (apomorphic) or ancestral (plesiomorphic). Similarly, if B and C are sister taxa composing a monophyletic group to the exclusion of A (panels *e* and *f* in Fig. A1), then wing presence in A and B could be due either to evolutionary convergence (panel *e*) or to ancestral character-state retention (panel *f*).

Among the six phylogenetic arrangements shown in Figure A1, wing presence is a shared derived (synapomorphic) character for particular ingroup taxa only in panel *a*. Thus, *if* wing presence is a genuine synapomorphy, then by Hennigian principles only panel *a* would provide a valid summary of the evolutionary history of these species and their characters. Otherwise, the shared presence of wings would be either inconclusive (panels *b*, *d*, and *f*) or positively misleading (panels *c* and *e*) with regard to identifying A and B as sister taxa in a clade. Hennig's seminal contributions to phylogenetic theory were to elaborate the then novel concept that

**Box A1. Phylogenetic terminology and concepts** (after Avise, 2004)

I. Classes of organismal resemblance
   (a) *phenetic similarity*: overall resemblance between organisms.
   (b) *patristic similarity*: the component of overall similarity due to shared ancestry.
   (c) *homoplastic similarity (homoplasy)*: in a narrow sense, the component of overall similarity due to evolutionary convergence from unrelated ancestors. The term is used also in a broad sense to mean all "extra evolutionary steps" in a phylogenetic tree, as could arise from convergence, parallelism, or evolutionary reversals in character states.

II. Classes of character state used to characterize organismal resemblance
   (a) *plesiomorphy*: an ancestral character state (i.e. one present in the common ancestor of the taxa under study).
   (b) *symplesiomorphy*: an ancestral character state shared by two or more descendant taxa.
   (c) *apomorphy*: a derived or newly evolved character state (i.e. one not present in the common ancestor of the taxa under study).
   (d) *synapomorphy*: a derived character state shared by two or more descendent taxa.
   (e) *autapomorphy*: a derived character state unique to a single taxon.

III. Other relevant definitions
   (a) *monophyletic group or clade*: an evolutionary assemblage that includes a common ancestor and all of its descendents; or, an assemblage that traces to a single node in an evolutionary tree.
   (b) *paraphyletic group*: an artificial assemblage that includes a common ancestor and some but not all of its descendents.
   (c) *polyphyletic group*: an artificial assemblage derived from two or more distinct ancestors; or, an assemblage that traces to two or more nodes in a phylogenetic tree.
   (d) *ingroup*: the species of phylogenetic interest (i.e. whose phylogeny is to be estimated).
   (e) *outgroup*: a taxon phylogenetically outside the clade of interest.
   (f) *sister taxa*: taxa stemming from the same node in a phylogeny.

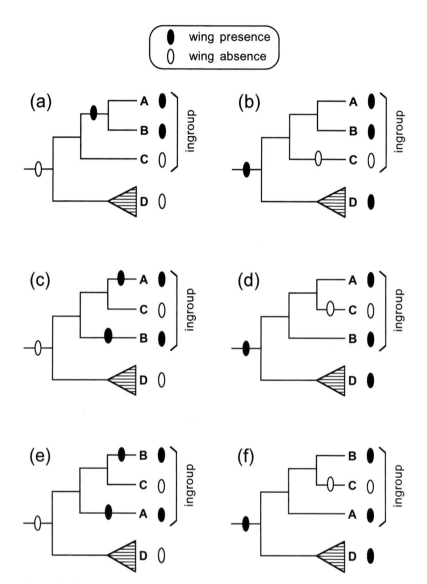

*Figure A1.* Hypothetical cladograms showing alternative phylogenetic arrangements for three ingroup taxa (A–C), plus an outgroup clade indicated by D (see text). Wing absence is the presumed ancestral condition in panels a, c, and e; wing presence is the ancestral condition in panels b, d, and f.

only valid synapomorphic characters can unambiguously earmark monophyletic groups, and, more generally, to promote the notion that various character-state distributions are related to phylogenetic patterns in logically interpretable ways.

Normally, Hennigian principles are employed when attempting to construct a cladogram from morphological or other phenotypic data (i.e. without having a molecular or other tree as independent phylogenetic backdrop). However, Hennigian logic is also highly relevant to PCM when a backdrop tree is available, as can be further illustrated by reference to Fig. A1. For the sake of argument, suppose now that we know (from extensive molecular genetic data or other evidence unrelated to wing presence versus absence *per se*) that the true phylogeny for the ingroup species is ((A, B) C). Then, by parsimony reasoning (see below), we can conclude that wings are either a synapomorphy (panel *a*) or a symplesiomorphy (panel *b*) for species A and B, and that evolutionary convergence was probably not involved. Furthermore, if we knew for certain that the ancestral character state was wing absence (panel *a*), we could then conclude that wings had most likely arisen along (and indeed cladistically define) the branch leading to the A + B clade.

In practice, how might that ancestral character state for the ingroup be inferred? Sometimes the answer is obvious. Suppose, for example, that A and B in Fig. A1 are two species of bat and that C and D are other mammals. Since nearly all mammals (as well as their reptilian ancestors) are flightless, the character state "wing presence" shared by A and B is clearly derived, so at this gross level of examination the two bats would belong to a clade distinct from C and D. However, much depends on the phylogenetic frame of reference, because what is a derived character state at one hierarchical level of the phylogeny might be an ancestral character state at another hierarchical level. Thus, A, B, C, and D might all be bat species, in which case the more immediate ancestor of A and B almost certainly had wings. In that refined context, wing presence in A and B would be a shared ancestral condition, and hence uninformative as to exactly how A and B might be related to one another *vis-à-vis* other bats.

In practice, the most common cladistic method for deducing the ancestral character state of an ingroup is to examine relevant outgroups as well, ideally including one that is a sister clade to the ingroup. The reasoning then proceeds as follows: a character state that is shared by the outgroup and some or all ingroup members was probably the ancestral (plesiomorphic) condition for the ingroup, such that other (derived) character states uniquely shared by particular ingroup members then become recognizable as putative synapomorphies that potentially define particular ingroup clades. Thus, panels *a*, *c*, and *e* in Fig. A1 suggest that the ancestral state for the ingroup was wing absence, in which case A and B would be deemed sister taxa earmarked by a synapomorphy (as in panel *a*); whereas panels *b*, *d*, and *f* suggest that the ancestral state for the ingroup was wing presence, in which case C would possess an autapomorphy and no ingroup clades are identified.

## Maximum parsimony

Hennig's reasoning provided a key conceptual advance in the elaboration of parsimony methods that attempt to account for character-state distributions in evolutionary ways that would seem to be as simple and straightforward as possible. A maximum parsimony (MP) reconstruction is one that requires the smallest number of evolutionary changes to explain the observed character-state differences among the particular taxa being compared. In other words, the most parsimonious tree minimizes the number of evolutionary steps required to account for a given set of empirical data. In general, the principle of parsimony is essentially Occam's razor as applied to phylogenetic issues.

For example, the cladogram in panel *a* of Fig. A1 has only one character-state change (one evolutionary step) in total along its branches, whereas the alternative trees in panels *c* and *e* have two character-step changes each. Thus, by parsimony criteria (and given that wing absence was the ancestral condition), tree *a* is to be preferred over trees *c* or *e*, each of which involves homoplasy. By similar logic, if we assume that wing presence was the ancestral condition (such that A and B display a symplesiomorphy), then the trees in panels *b*, *d*, and *f* involve only one character-state change each, and there is no basis for deciding among these phylogenies by the parsimony criterion.

None the less, even in this simplest possible case involving just three taxa and two alternative phenotypes (Fig. A1), countless other scenarios of character-state change are imaginable. In panel *c*, for example, the evolutionary lineage leading from the tree's root to extant species C could have experienced a gain and then a subsequent loss of wings, but then the overall tree would again have had a total of at least two character-state transitions. Indeed, in principle any number of back-and-forth steps between winged and wingless could have occurred along any tree branches in Fig. A1, but all such evolutionary interpretations would be less parsimonious than those already considered.

Such observations raise several fundamental and related issues about MP. First, a most parsimonious reconstruction for any real data is not necessarily a correct portrayal of the actual phylogenetic history of the species or characters analyzed, because evolution may follow winding paths rather than shortest direct routes. Second, different reconstructions of character-state transitions along a given phylogenetic tree can sometimes be equally parsimonious or nearly so (compare panels *a*, *b*, *d*, and *f*), in which case it can be impossible to determine without additional information which evolutionary reconstruction is correct (even assuming that evolution has taken the most parsimonious route). Third, when character-state changes occur rapidly along branches of a phylogenetic tree, any parsimony-based method of inference will probably fail to recover the true evolutionary history of those characters, at least to some extent. All of these points mean that MP in the context of PCM is best applied when character-state changes are relatively few and far between along the branches in a tree whose structure is known, and when additional information is available to help assess character-state polarities (ancestral versus derived conditions).

A fourth important point is that the "most parsimonious" phylogenetic explanation of data can depend critically on the evolutionary model assumed for character-state changes. For example, some character states might be freely reversible in evolution, whereas others may be biased in the direction of evolutionary transition or even non-revertible to particular ancestral states. To accommodate this possibility, some PCM computer programs permit users to specify the relative probabilities of different kinds of character-state transitions *a priori* (assuming that these are known or can be estimated from other lines of evidence). For example, a researcher might have genetic or ecological reasons to suspect that the evolutionary loss of wings is easier than wing gain, in which case a computer search for the most parsimonious PCM reconstruction(s) could be conducted accordingly. However, secure independent information about relative evolutionary transition rates is normally difficult to obtain. An extreme example of how alternative evolutionary models can affect PCM interpretations is detailed in the section *Dabbling into duck plumages* (Chapter 3).

In practice, PCM analyses typically involve many more taxa, more characters, and more evolutionary transitions between character states than are illustrated in Fig. A1. Software packages have been developed that assist in estimating ancestral character states and character-state changes, given a phylogenetic tree together with observed distributions of the relevant phenotypes in extant taxa. For PCM analyses via MP, the most widely used computer program is MacClade (Maddison and Maddison, 2000), which has user-friendly tutorial instructions (see also Hall, 2004) and excellent graphics. An example of the parsimony algorithms used by MacClade to reconstruct ancestral character states is detailed in Box A2.

The primary purpose and utility of MacClade is to reconstruct character-state evolution on any phylogenetic tree that is pre-specified by the user. For most of the case studies discussed in this book, such trees were estimated from molecular genetic data, and the authors then used MacClade (or similar programs) to interpret the evolutionary histories of specific phenotypes against that historical framework.

## Box A2. Parsimony reconstruction of ancestral character states
(after Cunningham *et al.*, 1998)

The data used to initiate the analysis consist of observed character states in extant species, such as the "black" versus "white" conditions in the five species shown in the diagrams below.

It is assumed that the correct phylogeny for those species is known from independent evidence.

In this example, it is assumed that the character states are equally weighted and initially unordered.

The algorithm in MacClade first conducts a "left-pass" traversal through the tree, using the following rules.

Rule 1.  If descendant nodes share any states, assign the set of shared states to the ancestor.

Rule 2.  If no states are shared in descendant nodes, assign an ambiguous state to the ancestor.

MacClade then conducts a "right-pass" traversal. In the first step of the right pass, the state of node d is determined by the left-pass state of node b and that of its sister node a; in the second step, the state of node e is determined by the state of node d and its sister node c; and so on.

MacClade then estimates a final optimization state for each ancestor, as illustrated for node b as follows. It jointly considers the right-pass state of that node, and the left-pass states of its two descendant nodes, and then chooses the state with the greatest representation among those three sets. If no state is in the majority, the node remains ambiguous.

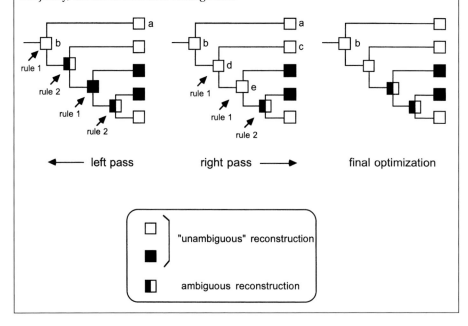

There is no reason why phenotypic data themselves could not also serve as the basis of the specified tree, but ideally those phenotypic data should be unrelated to the character states subsequently being mapped (lest there be a potential for circular reasoning).

Until very recently, MP was the only formal method available for mapping phenotypic character states onto phylogenies (Huelsenbeck *et al.*, 2003). Although MP has many complexities and potential pitfalls, most biologists agree that it is probably better to use parsimony even crudely than to ignore phylogenetic history altogether in attempts to understand the evolutionary histories of character-state transitions. The bottom line, however, is that a highly conservative stance should normally be adopted when interpreting results of PCM under parsimony (or under any other PCM reconstruction method).

## Maximum likelihood

A typical PCM reconstruction based on MP is summarized as a cladogram with particular ancestral character states provisionally specified at various internal nodes and along tree branches. At face value, any such representation has a shortcoming in that it fails to rule out alternative reconstructions also consistent with the data, such that uncertainty in the evolutionary estimate is not formally quantified. A more recent approach to PCM analysis – involving maximum likelihood (ML) – partly alleviates this problem by explicitly addressing the relative probabilities of alternative character states inside a tree.

In general, ML methods in phylogenetics operate under the principle that reconstructions should maximize the probability of observing an available data set given a particular model of evolutionary change (Kolaczkowski and Thornton, 2004). An explicit evolutionary model (which itself is often informed by the empirical data available in a particular study) is needed to calculate the relative likelihoods of different character states at internal nodes. A typical model might assume, for example, that: (a) the probability of evolutionary change at any point in time along any branch of the tree depends only on the character state at that time (and not on prior states); (b) evolutionary transitions along each branch are independent of changes elsewhere in the tree; and (c) rates of change between any two character states (the rates in opposite directions may either be equal or unequal) are constant throughout all branches of the tree. Thus, another feature of ML approaches to PCM is that branch lengths (not merely cladistic branching order) can be taken into account in reconstructing character states. A typical ML output gives relative probabilities (often shown as pie diagrams) of alternative character states at internal nodes in the tree.

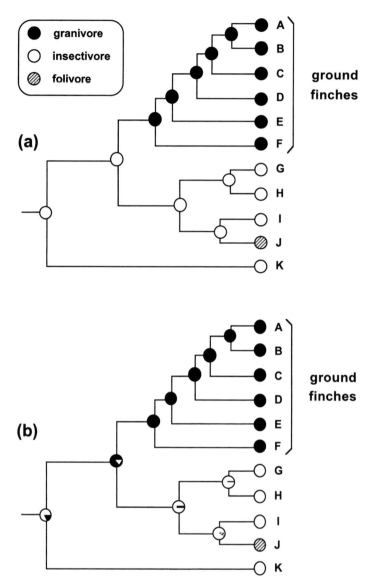

*Figure A2.* Results of MP (a) and ML (b) PCM analyses of dietary habits in the evolutionary history of 11 species of Galápagos finch (after Schluter *et al.*, 1997). Areas in the pie diagrams of (b) indicate relative levels of support for different ancestral states.

The following PCM example (adapted from Schluter *et al.*, 1997) will illustrate a ML output and its comparison to a MP output for the same set of data. The Galápagos finches include about a dozen living bird species that are the result of a recent adaptive evolutionary radiation on the Galápagos archipelago. Collectively, they display three distinct dietary habits: granivory (seed eating), insectivory, and folivory (foliage eating). For these dietary regimes, Fig. A2 summarizes PCM reconstructions based on MP (above) and ML (below). The phylogenetic tree itself had been estimated from molecular genetic data, and the PCM analysis began with the known dietary habits of 11 extant finch species (A–K).

In this case, the two methods of PCM analysis yielded similar evolutionary reconstructions (as would be hoped), but they also yielded some notable differences (Fig. A2). First, whereas the MP output portrays the evolutionary history of dietary changes as unambiguous, the ML output shows inherent uncertainties at particular interior nodes. Second, whereas the MP analysis shows insectivory as the ancestral state in the common ancestor of extant species A–J, the ML analysis suggests that granivory was the more probable ancestral condition for that clade. This latter discrepancy arises because the ML analysis takes tree branch lengths into account (whereas the MP analysis does not), and less evolutionary time was available for shifts in diet along the phylogenetic branch leading to the ground finches.

By examining several such case studies involving a variety of taxa and different categories of phenotypes, Schluter *et al.* (1997) concluded that MP and ML analyses generally yield similar PCM reconstructions when character-state changes are relatively rare on a phylogenetic tree, but that outputs of all PCM reconstruction methods become much less certain when character-state changes have occurred far more frequently in evolution. In such cases, it may thus be necessary (as well as sufficient for some interpretive purposes) merely to conclude from PCM analyses that many character-state transitions have occurred within a particular phylogenetic tree. In other words, a conservative interpretation would not go too far in detailing precise numbers or exact phylogenetic locations of particular types of character-state transitions.

## Independent contrasts between pairs of quantitative traits

In comparative evolutionary research, a common approach is to examine joint distributions of two or more phenotypic traits across a range of species or higher taxa, one objective being to identify correlational patterns that might imply causal relationships. However, even functionally unrelated phenotypes can co-vary across taxa simply by virtue of historical associations with one another. Accordingly, many biologists argue that phylogeny should always be taken into account when evaluating character-state correlations in a comparative context. Although counterarguments to this wisdom sometimes are expressed (see Ricklefs, 1996; Price, 1997), it remains true that egregious errors in data interpretation can arise when phylogenetic considerations are neglected.

Consider, for example, a hypothetical case in which body size (large or small) and extinction risk (high versus low) co-vary perfectly across 12 species (Fig. A3, top). The interpretation of this obvious association can differ greatly depending on the species' phylogeny. If these species are phylogenetically related as shown on the left side of Fig. A3, then the character correlation is statistically and evolutionarily significant because, in each of six phylogenetically independent contrasts between large-bodied and small-bodied species, largeness was invariably associated with high extinction risk. Thus, it might be quite appropriate to conclude that large-bodied species inherently tend to be more vulnerable to extinction than small-bodied species (perhaps in this case because a third variable such as small population size might be biologically associated with both high extinction risk and large bodies). On the other hand, if these species are phylogenetically related as shown on the right side of Figure A3, then the character correlation is statistically non-significant (after correcting for phylogeny) because there might have been only one evolutionary event (at point X) that eventuated in the perfect association of large body with high extinction risk in extant species. If so, it would be premature to conclude that large-bodied species are inherently more vulnerable to extinction than small-bodied species (except insofar as this might be so because of phylogenetic legacies).

Figure A4 illustrates another way to consider this potential problem of history-based non-independence between pairs of characters in a quantitative data set. Suppose that quantitative traits X and Y have both been measured in each of six extant species whose phylogeny is known with certainty (upper diagram in Fig. A4). For example, trait X might again be body size and trait Y might be extinction risk; or X and Y could be any other characters for which numerical values are available for each species. The null hypothesis to be tested is that these traits do not evolutionarily co-vary across these species. If all six species are viewed as providing statistically independent data points in a two-axis graph between traits X and Y, then, in principle, a phylogenetically naïve researcher could be seriously misled in either of two ways depending on the particular trait values observed. On the one hand (lower left diagram of Fig. A4), that researcher might erroneously reject the null hypothesis when in fact it is true. On the other hand (lower right diagram of Fig. A4), (s)he might mistakenly accept the null hypothesis when actually it is false. Thus, statistical errors of both Type I (rejection of a null model that is true) and Type II (acceptance of a null model that is false) could be imagined if phylogenetic corrections were not properly incorporated into the analysis.

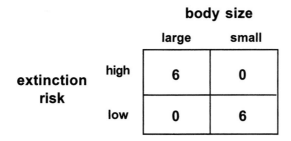

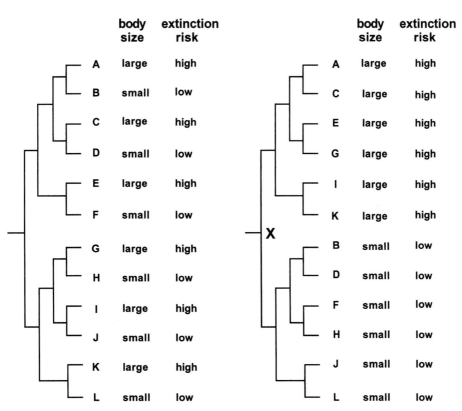

*Figure A3.* Illustration of how evolutionary history can affect conclusions about character-state correlations (after Fisher and Owens, 2004). In both of the phylogenetic diagrams, body size and extinction risk are perfectly correlated among the 12 extant species (A–L), but the statistical and biological interpretations of these data greatly depend on how these species are phylogenetically related (see text).

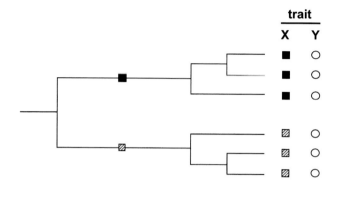

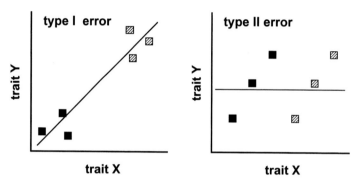

*Figure A4.* Above: phylogenetic relationships among six extant species for which quantitative measurements for traits X and Y are to be compared. For simplicity, only low versus high values for X are distinguished (by filled and hatched squares, respectively). Below: bivariate plots of hypothetical values for traits X and Y across these six species. In the graph on the left, a strong correlation exists between species trait values for X and Y, but this could be biologically "spurious" owing to phylogenetic constraints imposed by the ancestral conditions (as shown in the phylogeny above). In the graph on the right, there is no overall correlation between the values for traits X and Y, but this too could be misleading because biologically meaningful associations could exist in subsets of the phylogeny. See text for additional explanation. This figure is modified, with permission, from drawings by Rich Grenyer (based on a treatment by Harvey and Pagel, 1991).

The method of independent contrasts (Felsenstein, 1985; Garland *et al.*, 1992) is one of several approaches by which PCM analyses can be statistically corrected for phylogenetic non-independencies of the sort illustrated in Figs. A3 and A4 (Harvey *et al.*, 1996; Martins, 1996). A popular computer program for implementing these analyses is CAIC (Comparative Analysis by Independent Contrasts) by Purvis and Rambaut (1995). The basic idea, elaborated in Box A3, is to avoid comparisons

## Box A3.  Trait comparisons by independent contrasts

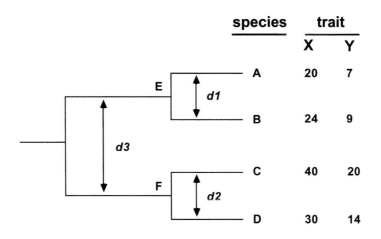

| species | trait | |
|---|---|---|
| | **X** | **Y** |
| A | 20 | 7 |
| B | 24 | 9 |
| C | 40 | 20 |
| D | 30 | 14 |

Consider the phylogeny shown in the figure above, and assume that, for each of the extant species (A–D), traits X and Y have been measured or quantified. To correct for phylogenetic non-independence among these species' trait values, "independent contrasts" are first identified. There are three such phylogenetically independent contrasts for the four species depicted: $d1$ (the difference between a trait's values in sister species A and B); $d2$ (the difference between a trait's values in sister species C and D; and $d3$ (the difference between a trait's values at interior nodes E and F). For example, $d1$ and $d2$ for trait X are 4 and 10, respectively. To calculate $d3$, a trait's values at the internal nodes E and F must first be deduced. These can be estimated as the mean of the trait values in the clades they earmark, such that the deduced values for trait X in E and in F are 22 and 35, respectively. Thus, $d3$ for trait X is 13. Similar logic applies to the calculations of $d1$, $d2$, and $d3$ for trait Y. The overall result is a table of independent contrasts (below) for traits X and Y.

| | | X | Y |
|---|---|---|---|
| **table of** | $d1$ | 4 | 2 |
| **independent contrasts** | $d2$ | 10 | 6 |
| | $d3$ | 13 | 9 |

These independent contrasts then are graphed on a bivariate plot (below) that can reveal whether of not significant evolutionary associations exist (after accounting for phylogeny) between traits X and Y. (These figures are modified, with permission, from drawings by Rich Grenyer (based on a treatment by Harvey and Pagel, 1991).)

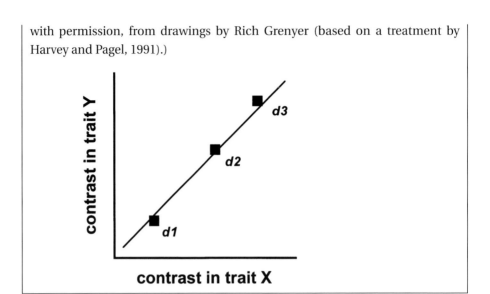

that involve overlapping branches (historical redundancies) in the phylogenetic trees of the species examined. By hard criteria, only after such historical non-independence among lineages has been statistically accommodated does it then become appropriate to consider possible biological reasons for any genuine evolutionary correlations among traits that might be detected.

# GLOSSARY

**Abdomen**  In mammals, the body region between the diaphragm and pelvis; in arthropods, the hind-body.

**Adaptation**  Any feature (e.g. morphological, physiological, behavioral) that helps an organism to survive and reproduce in a particular environment.

**Adaptive radiation**  An evolutionarily rapid proliferation of related species as they diversify to occupy varied environmental conditions.

**Alga (pl. algae)**  Any of an assemblage of photosynthetic non-vascular organisms, which differ from more advanced plants in several features including their lack of multicellular sexual organs.

**Allele**  Any of the possible alternative forms of a gene. A diploid individual carries two alleles at each autosomal gene; these can either be identical in state (in which case the individual is homozygous) or different in state (heterozygous). At each autosomal gene, a population of $N$ diploid individuals carries $2N$ alleles, many of which may differ in details of nucleotide sequence.

**Allen's rule**  The generality that homeothermic animals tend to have shorter appendages in colder climates.

**Alloparental care**  The rearing of foster children.

**Allopatric**  Inhabiting different geographic areas.

**Altruism**  Selfless behavior performed for the benefit of others.

**Amino acid**  One of the molecular subunits that when joined together form a polypeptide.

**Anagenesis**  Genetic change through time within a single evolutionary lineage.

**Analogous characters**  Traits that have similar functions but different evolutionary origins.

**Anther**  The pollen-bearing part of a flower's stamen.

**Anthropoid**  Of or pertaining to human-like apes such as gorillas and chimpanzees, or more generally to all higher primates.

**Anti-tropical**  Occurring at mid or high latitudes in both the Northern and Southern Hemispheres, but not in between.

**Apomorphy**  A derived or newly evolved character state (i.e. one not present in the common ancestor of the taxa under study).

**Aposematic coloration**  *See* **warning coloration**.

**Area cladogram**   A branching diagram summarizing the historical relationships of particular geographic regions.

**Asexual reproduction**   Any form of reproduction that does not involve the fusion of sex cells (gametes).

**Autapomorphy**   A derived character state unique to a single taxon.

**Autosome**   A chromosome in the nucleus other than a sex chromosome. In diploid organisms, autosomes are present in homologous pairs.

**Bacterium** (pl. **bacteria**)   A unicellular micro-organism without a true cellular nucleus.

**Behavior**   Any observable action or response of an organism.

**Bergmann's rule**   The generality that homeothermic animals tend to have larger body sizes (lower surface : volume ratios) in colder climates.

**Bilateral symmetry**   A body plan in which an organism can be divided by a longitudinal plane into two parts that are mirror images of each other.

**Biodiversity**   Life's genetic heterogeneity, at any or all levels of biological organization.

**Biological species**   *See* **species**.

**Biotype**   A recognizable biological form, usually in reference to asexual or parthenogenetic taxa for which the term "species" does not always apply well.

**Branch**   An extended ancestral–descendent lineage between nodes in a phylogenetic tree.

**Broadcast spawning**   The release of gametes or larvae into the open water during the reproductive process.

**Brood**   *n.*, a parent's group of eggs or offspring; *v.*, to tend or rear such a group.

**Brood parasitism**   The use of a foster parent to raise an individual's young. The foster parent may be of the same species (intraspecific brood parasitism) or of a different species (interspecific brood parasitism).

**Brood pouch**   An anatomical bag or biological sac for housing developing offspring.

**Carnivorous**   Flesh-eating.

**Cell**   A small, membrane-bound unit of life capable of self-reproduction.

**Character state**   A phenotypic condition recognizably distinct from others of similar type.

**Chloroplast**   An organelle in the cytoplasm of plant cells that contains its own DNA (cpDNA) and is the site of photosynthesis.

**Chromosome**   A threadlike structure within a cell that carries genes.

**Clade**   A group of species (or, sometimes, individuals) that share a closer common ancestry with one another than with any other such group; a monophyletic assemblage.

**Cladistics**   *See* **Hennigian cladistics**.

**Cladogenesis**   The splitting or branching of evolutionary lineages; normally equated with speciation.

**Cladogram**   A diagram showing the branching topology (but not necessarily branch lengths) in an evolutionary tree.

**Classification**   A process of establishing, defining, and ranking biological taxa within hierarchical groups; or, the outcome itself of this process.

**Class, taxonomic**   A hierarchical rank between phylum and order in a traditional classification scheme.

**Clone**   *n.*, a group of genetically identical cells or organisms, all descended from a single ancestral cell or organism; *v.*, to produce such genetically identical cells or organisms.

**Clutch**   *See* **brood** (*n.*).

**Coalescent theory**   The body of mathematical thought concerning how alleles in a population trace back through the pedigree to common ancestral states in the past.

**Co-evolution**   The joint evolution of two of more ecologically interacting species.

**Congeneric**   Belonging to the same genus.

**Conservation biology**   The theory and practice of protecting biodiversity.

**Conspecific**   Belonging to the same species.

**Continental drift**   The movement of continents across the earth's surface over geological time.

**Convergent evolution**   The independent evolution of structural, functional, or other similarities between distantly related or unrelated species.

**Countercurrent heat exchanger**   Usually a system of closely oppressed arteries and veins, in a particular body region, that serves to help maintain internally generated body heat.

**Cryptic species**   Reproductively isolated species that resemble one another closely and, thus, sometimes remain unrecognized.

**Cuckold**   An individual whose mate has been unfaithful.

**Cuckoldry**   The act of making a cuckold.

**Cytoplasm**   The portion of a cell outside of the nucleus.

**Cytoplasmic genome**   A genome housed within the cytoplasm of a eukaryotic cell.

**Delayed implantation**   A phenomenon in which an assemblage of postzygotic cells in a pregnant female is developmentally arrested for an extended period of time before implanting into the uterine wall, where embryonic development then resumes.

**Deoxyribonucleic acid (DNA)**   The genetic material of most life forms; a double-stranded molecule composed of strings of nucleotides.

**Diploid**   A usual condition of a somatic cell in which two copies of each chromosome are present.

**Dispersal**   Any spatial movement of an individual, usually away from its birth site or prior breeding location.

**DNA–DNA hybridization**   A class of laboratory procedures that measures the extent to which single-strand stretches of polynucleotides attract and bind to homologous, complementary single strands.

**DNA repair**   Mending of DNA damage, done naturally in cells by special enzymatic apparatuses.

**DNA sequencing**   Any laboratory procedure by which the sequence of nucleotides in a nucleic acid is determined.

**Dollo's law**   The notion that a complex adaptation, once lost, can never be regained in exactly the same form.

**Domain, taxonomic**   A hierarchical rank above kingdom in a classification scheme.

**Dominant allele**   A form of a gene that masks the phenotypic expression of its recessive counterpart.

**Echolocation**   The use of high-frequency sound waves to perceive physical objects in the environment.

**Ecogeographic rule**   A tendency for particular adaptive phenotypes to appear in particular ecological or geographical settings. *See also* **Allen's rule**, **Bergmann's rule**, and **Gloger's rule**.

**Ecology**   The study of the interrelationships among living organisms and their environments.

**Ecomorph**   A recognizable morphological type within a species, usually associated with a particular ecological setting.

**Ecosystem**   A community of ecologically interacting organisms and their environment.

**Ectothermic**   Having a body temperature determined primarily by the temperature of the environment; "cold-blooded."

**Egg**   A female gamete.

**Egg dumping**   *See* **brood parasitism**.

**Embryo**   A developing organism between fertilization and birth or hatching.

**Embryonic diapause**   A temporary cessation of embryonic development.

**Endangered species**   A species at immediate risk of extinction.

**Endemic**   Native to, and restricted to, a particular geographic area.

**Endosymbiotic theory**   A hypothesis stating that, early in the history of life, amalgamations of prokaryotic microbes eventuated in eukaryotic cells with distinctive nuclear and cytoplasmic genomes.

**Endothermic**   Maintaining a body temperature largely independent of the temperature of the environment; "warm-blooded."

**Enzyme**   A protein that catalyzes a specific chemical reaction.

**Epidemiology**   The study of disease outbreaks, including an effort to trace possible causes.

**Ethology**   Behavior.

**Ethotype**   A recognizable behavioral form within a species.

**Eukaryote**   Any organism in which chromosomes are housed within a membrane-bound nucleus.

**Eusociality**   A social system characterized by cooperative care of young and reproductive division of labor, with sterile individuals working on behalf of reproducers within a colony.

**Evolution**   Any change across time in the genetic composition of populations or species.

**Evolutionary plasticity**    Of or pertaining to organismal features that can change rapidly or fluidly during the evolutionary process.

**Evolutionary tree**    *See* **phylogeny**.

**Exon**    A coding segment of a gene. *See also* **intron**.

**Exoskeleton**    An external covering or integument, usually hard.

**Exotic**    Not native to the geographic area in question.

**Extinction**    The permanent disappearance of a population or species.

**Family, taxonomic**    A hierarchical rank between order and genus in a traditional classification scheme.

**Fecundity**    The potential reproductive capacity of an individual, usually measured as the number of gametes produced.

**Fermentation**    An enzymatically controlled anaerobic breakdown of organic materials.

**Fertilization**    The union of two gametes to produce a zygote.

**Fitness (genetic)**    The contribution of an individual (or of a particular genotype) to the next generation, relative to the contributions of other individuals (or genotypes) in the population.

**Foregut**    A front section in the digestive tract.

**Forensics (genetic)**    Of or pertaining to the diagnosis of otherwise unknown biological material based on analysis of proteins or DNA.

**Fossil**    Any remains or trace of past life.

**Foster parentage**    The rearing of young that are not one's own biological offspring.

**Founder effect**    Genetic consequences that follow the establishment of a new population from a small number of colonizing individuals.

**Frequency-dependent selection**    A form of natural selection that operates differently depending on genotypic or phenotypic frequencies within a population. For example, natural selection might disproportionately favor particular traits when they are rare, in which case a balanced polymorphism could be maintained in a population.

**Fungus (pl. fungi)**    Any of an assemblage of non-vascular, non-photosynthetic eukaryotic organisms that includes such diverse forms as molds, yeasts, rusts, and mushrooms.

**Gamete**    A mature reproductive sex cell (egg or sperm).

**Gene**    The basic unit of heredity; usually taken to imply a sequence of nucleotides specifying production of a polypeptide or other functional product, but also can be applied to stretches of DNA with unknown or unspecified function.

**Genealogy**    A record of descent from ancestors through a pedigree.

**Gene flow**    The spatial movement of genes, normally within a species.

**Gene pool**    The sum total of all hereditary material in a population or species.

**Genetic drift**    Any shift in allele frequency in a population by chance sampling of gametes from generation to generation.

**Genetic engineering**    The purposeful alteration of genetic material by humans.

**Genetic markers**    Natural DNA (or RNA) tags that exist in all forms of life.

**Gene tree**    A diagram of phylogenetic relationships among alleles at any specified locus. *See also* **species tree**.

**Genome**    The complete genetic constitution of an organism; also can refer to a particular composite piece of DNA, such as the mitochondrial genome.

**Genomics**    The study of genomes.

**Genotype**    The genetic constitution of an individual with reference to a single gene or set of genes.

**Genus, taxonomic**    A hierarchical rank between family and species in a traditional classification scheme.

**Germ cell**    A sex cell or gamete.

**Gloger's rule**    The generality that animals tend to have darker pigmentation in geographical areas with high humidity.

**Gondwanaland**    The supercontinent of the Southern Hemisphere, in existence more than 150 million years ago before Africa, South America, India, Australia, and Antarctica began to move apart by continental drift.

**Haploid**    A usual condition of a gametic cell in which one copy of each chromosome is present.

**Hennigian cladistics**    The study of branching relationships of lineages in an evolutionary tree.

**Herbivorous**    Plant-eating.

**Heredity**    Inheritance of genes; the phenomenon of familial transmission of genetic material from one generation to the next.

**Hermaphrodite**    A condition in which an individual produces both male and female gametes. If both types of gamete are generated at the same life stage, the individual is a synchronous hermaphrodite. If they are generated sequentially during an organism's lifetime, the individual is either a protandrous hermaphrodite (male gametes first) or a protogynous hermaphrodite (female gametes first).

**Heterochrony**    An evolution change in the onset of a developmental process in individuals, or in their particular organs or other features.

**Heterogametic sex**    The gender that produces gametes containing unlike sex chromosomes.

**Hibernation**    The act or condition of being in a dormant or resting state during the winter.

**Homeothermic**    *See* **endothermic**.

**Homeotic gene**    A gene with profound morphotypic influences during individual development.

**Homogametic sex**    The gender that produces gametes containing alike sex chromosomes.

**Homology**    Similarity of features (morphological, molecular, etc.) due to inheritance from a shared ancestor.

**Homoplasy**   The component of phenotypic similarity not due to shared ancestry (and therefore due to other processes such as evolutionary convergences or reversals of character states).

**Horizontal gene transfer**   The movement of genetic material between organisms by means other than normal vertical transmission from parents to offspring.

**Hormone**   A chemical substance, secreted naturally by endocrine glands, that produces a specific physiological response in a distant target tissue.

**Hybridization**   The successful mating of individuals belonging to genetically different populations or species.

**Ice Ages**   Times of climatic cooling and formation of vast continental glaciers at high latitudes, as occurred repeatedly during the Pleistocene epoch.

**Implantation**   The attachment of an embryo to the uterine wall.

**Inclusive fitness**   An individual's own genetic fitness as well as his or her effects on the genetic fitness of close relatives. *See also* **fitness**.

**Ingroup taxa**   Species of phylogenetic interest (i.e. whose phylogeny is to be estimated).

**Introgression**   The movement of genes between species via hybridization and back-crossing.

**Intron**   A non-coding portion of a structural gene. Most such protein-coding genes in eukaryotic organisms consist of alternating intron and exon sequences.

**Invertebrate**   An animal that does not possess a backbone.

**Jumping gene**   *See* **transposable element**.

**Junk DNA**   A term often used to describe nucleic acid sequences that do not actively contribute to cellular functions such as by specifying an operational protein or RNA product. *See also* **selfish DNA**.

**Key evolutionary innovation**   A newly evolved trait that predisposes a genetic lineage to an adaptive radiation.

**Kingdom, taxonomic**   A hierarchical rank above phylum in a traditional classification scheme.

**Kin selection**   A form of natural selection due to individuals favoring the survival and reproduction of genetic relatives (other than their own offspring).

**Larva** (pl. **larvae**)   The distinctive, pre-adult form in which some animals hatch from the egg.

**Lecithotrophy**   A developmental pattern in which embryos feed upon egg yolk. *See also* **matrotrophy**.

**Lichen**   An amalgamation involving the symbiotic association between an alga and a fungus.

**Life cycle**   The sequence of events for an individual, from its origin as a zygote to its death; one generation.

**Live-bearing**   Viviparity. *See* **viviparous**.

**Locus** (pl. **loci**)   A gene, or a specified sequence region of DNA.

**Macroevolution**   Genetic differentiation through time among species and higher taxa.

**Magnetotaxis**   An organism's ability to sense and orient to a magnetic field.

**Marsupium**   A brood pouch.

**Mating system**   The particular pattern by which males and females, or their gametes, come together during the reproductive process. *See also* **monogamy, polyandry, polygamy, polygynandry,** and **polygyny**.

**Matriline**   A genetic transmission pathway strictly through females (as traversed, for example by animal mtDNA).

**Matrotrophy**   A developmental pattern in which embryos receive food directly from their mothers. *See also* **lecithotrophy**.

**Maximum parsimony**   *See* **parsimony**.

**Meiosis**   The cellular process whereby a diploid cell divides to form haploid gametes.

**Melanistic**   Darkly pigmented.

**Mesozoic Era**   The geological time frame beginning about 250 million years ago and ending about 65 million years before the present; the "age of the dinosaurs."

**Metabolism**   The sum of all physical and chemical processes by which living matter is produced and maintained, and by which cellular energy is made available.

**Metazoan**   A multicellular animal.

**Microbe**   A very small organism visible only under a microscope.

**Microevolution**   Genetic changes through time within a species.

**Migration**   A periodic, typically seasonal movement to and from a given geographical area, often along a consistent route.

**Mimicry**   The evolution of a close resemblance between any two unrelated species to deceive a third.

**Mitochondrion**   An organelle in the cytoplasm of animal and plant cells that contains its own DNA (mtDNA) and is the site of some of the primary metabolic pathways involved in cellular energy production.

**Mobile element**   *See* **transposable element**.

**Molecular clock**   An evolutionary timepiece based on the evidence that genes or proteins tend to accumulate mutational differences at roughly constant rates in particular sets of lineages.

**Molecular markers**   *See* **genetic markers**.

**Molecular phylogeny**   An evolutionary tree estimated from nucleic acid or protein information.

**Monogamy**   A mating system in which each male is paired with only one female, and vice versa.

**Monophyletic**   In evolution, tracing to a common or shared ancestor.

**Morphology**   The visible structures of organisms.

**Mosaic evolution**   Variation in rates or patterns of evolution in different kinds of traits.

**Müllerian mimicry**   Imitative similarity in warning displays (such as body color patterns) by two or more potential prey species that are unpalatable or otherwise offensive to predators.

**Mutation**   A change in the genetic constitution of an organism or its gametes.

**Mutualism**   A form of symbiosis in which both organisms benefit from the association.

**Natural history**   The study of nature and natural phenomena.

**Naturalist**   A person interested in natural history.

**Natural selection**   The differential contribution by individuals of different genotypes to the population of offspring in the next generation.

**Neotropical**   Of or pertaining to regions in the American tropics.

**Nepotism**   Favoritism directed toward one's genetic kin.

**Neurotransmitter**   Any of several chemical substances that transmit nerve impulses between cells of the nervous system.

**Node**   A branching point within a phylogenetic tree (interior node), or the current tip of an outer branch (exterior node).

**Nuclear DNA**   Genetic material housed within the nucleus of eukaryotic cells.

**Nucleic acid**   *See* **deoxyribonucleic acid**.

**Nucleotide**   A unit of DNA consisting of a nitrogenous base, a pentose sugar, and a phosphate group.

**Nucleus** (pl. **nuclei**)   The portion of a cell bounded by a nuclear membrane and containing chromosomes.

**Ontogeny**   The course of development and growth of an individual to maturity.

**Order, taxonomic**   A hierarchical rank between class and family in a traditional classification scheme.

**Organelle**   A complex, recognizable structure in the cell cytoplasm (such as a mitochondrion or chloroplast).

**Outgroup taxa**   One or more species phylogenetically outside, but close to, the clade of interest.

**Oviparous**   Egg-laying.

**Paedomorphosis**   An evolutionary phenomenon in which adult descendents resemble the youthful stage of their ancestors.

**Palearctic**   Of or pertaining to high-latitude regions of Eurasia.

**Paleontology**   The study of extinct forms of life, normally through fossils.

**Paraphyletic**   An artificial assemblage that includes a common ancestor and some but not all of its evolutionary descendents.

**Parasite**   An organism that at some time in its life cycle is intimately associated with and harmful to a host.

**Parsimony**   Economy or sparingness of explanation; in phylogeny, the simplest set of evolutionary pathways to account for observed differences between taxa.

**Parthenogenesis**   The development of an individual from an unfertilized egg.

**Parturition**   The act of giving birth.

**Pathogen**   An organism that produces a disease.

**Patristic similarity**   The component of phenotypic similarity due to shared ancestry.

**Pedigree**   A diagram displaying population ancestry (mating partners and their offspring across generations).

**Phenetic similarity**   The overall phenotypic resemblance between any specified organisms.

**Phenogram**   A tree-like diagram summarizing phenetic similarities between organisms.

**Phenotype**   The observable properties of an organism at any level, ranging from molecular and physiological to anatomical and behavioral.

**Phenotypic plasticity**   The capacity of different phenotypes to emerge when an organism is exposed to different environmental conditions.

**Pheromone**   A chemical message, secreted by an individual, that conveys information to and often elicits a specific response from another individual.

**Photosynthesis**   The biochemical process by which a plant uses light to produce carbohydrates from carbon dioxide and water.

**Phylogenetic**   Of or pertaining to phylogeny.

**Phylogenetic character mapping (PCM)**   The science of deducing the evolutionary histories of phenotypic characters by plotting the distributions of their alternative states on phylogenetic trees.

**Phylogenetic constraint**   Evolutionary limitations on phenotypes due to historical legacies.

**Phylogenetic inertia**   The evolutionary maintenance or continuance of particular phenotypes due to genetic restraints imposed by history.

**Phylogenetic legacy**   That which is bequeathed by evolutionary history.

**Phylogenetic reconstruction**   The scientific practice of deducing the history and evolutionary relationships of genetic lineages.

**Phylogenetic tree**   *See* **phylogeny**.

**Phylogeny**   Evolutionary relationships (historical descent) of a group of organisms or species.

**Phylogeography**   A scientific field concerned with the spatial distributions of genealogical lineages, including those within species.

**Phylogram**   A diagram showing both branching topology and branch lengths in an evolutionary tree.

**Phylum, taxonomic**   A hierarchical rank between kingdom and class in a traditional classification scheme.

**Physiology**   The scientific study of tissue operations and metabolic functions of living organisms.

**Pistil**   The seed-bearing or ovule-bearing female organ of a flower.

**Placenta**   A physical structure that connects an unborn animal to its parent.

**Plankton**   Small organisms that float freely in the ocean or other bodies of water.

**Planktotrophy**   A life history that includes a plankton-feeding stage.

**Plate tectonics**   The concept that earth's crust is divided into a series of rigid panels or plates that slowly move relative to one another.

**Plesiomorphy**    An ancestral character state (i.e. one present in the common ancestor of the taxa under study).

**Pleistocene Epoch**    The geological time frame beginning about two million years ago and ending about ten thousand years before the present.

**Poecilogony**    A situation in which lecithotrophy and planktotrophy co-occur as life-history alternatives within a species.

**Poikilothermic**    *See* **exothermic**.

**Pollen**    A male gamete in plants.

**Pollination**    The transfer of pollen to a female flower or flower part.

**Polyandry**    A mating system in which a female acquires and mates with multiple males, but a male typically has only one mate at most. *See also* **polygyny** and **polygamy**.

**Polygamy**    A mating system in which an individual has more than one mate. *See also* **polygyny** and **polyandry**.

**Polygynandry**    A mating system in which both males and females normally have several mates.

**Polygyny**    A mating system in which a male acquires and mates with multiple females, but a female typically has only one mate at most. *See also* **polyandry** and **polygamy**.

**Polymerase chain reaction (PCR)**    A laboratory procedure for the replication of DNA *in vitro* from even small quantities of starting material.

**Polymorphism**    With respect to particular genotypes or phenotypes, the presence of two or more distinct forms in a population.

**Polypeptide**    A string of amino acids.

**Polyphyletic**    A group of organisms perhaps classified together, but tracing to different ancestors.

**Population**    Conspecific individuals inhabiting a defined area or sharing in a common gene pool.

**Population bottleneck**    A severe but temporary reduction in the size of a population.

**Population structure (genetic)**    Differences in genetic make-up among geographic populations.

**Predator**    An organism that feeds by preying on other organisms.

**Pregnancy**    Gestation of an embryo within a parent's body.

**Progenitor**    An ancestor.

**Prokaryote**    Any organism that lacks a chromosome-containing, membrane-bound nucleus.

**Protein**    A macromolecule composed of one or more polypeptide chains.

**Qualitative character**    A character for which alternative distinct states can be distinguished.

**Quantitative character**    A phenotypic character that varies more or less continuously among the taxa compared; may also refer to a phenotypic character with a complex or multi-factorial genetic basis.

**Radial symmetry**    A body plan in which an organism roughly resembles a circle or cylinder.

**Recapitulation**   An evolutionary phenomenon in which the juvenile stage of descendents resembles the adult stage of their ancestors.

**Recessive allele**   A form of a gene whose phenotypic expression is masked by its dominant counterpart.

**Recombinant DNA techniques**   Laboratory methods by which DNA sequences from different organisms are isolated and then spliced together in new arrangements.

**Recombination (genetic)**   The formation of new combinations of genes, as for example occurs naturally via meiosis and fertilization.

**Regulatory gene**   A gene that exerts operational control over the expression of other genes.

**Reproductive isolation**   Barriers to successful hybridization or introgression between biological species.

**Reticulate evolution**   Lateral transfer of genetic material between lineages, as mediated for example by introgressive hybridization or by one or another mechanism of horizontal gene transfer, and producing anastomotic connections between branches in a phylogenetic tree.

**Reticulation event**   An occurrence that results in reticulate evolution.

**Retroposon**   *See* **retrotransposable element**.

**Retrotransposable element**   A form of jumping gene or mobile element that transposes via an RNA intermediate.

**Retrovirus**   Any RNA virus that utilizes reverse transcription during its life cycle to integrate into the DNA of host cells.

**Ribonucleic acid (RNA)**   The genetic material of many viruses, similar in structure to DNA. Also, any of a class of molecules that normally arise in cells from the transcription of DNA.

**Ribosomal DNA**   Genetic material that encodes ribosomal subunits.

**Ribosome**   A cytoplasmic organelle that is the site of protein translation (i.e. where RNA is "read" by the cell to produce polypeptides).

**Root**   The most basal branch (pre-dating the earliest node) in a phylogenetic tree.

**Saprobe**   An organism that acts as a decomposer by absorbing nutrients from dead organic matter.

**Selfish DNA**   DNA that displays self-perpetuating modes of behavior without apparent benefit to the organism. *See also* **junk DNA**.

**Sex**   The male or female gender.

**Sex chromosome**   A chromosome in the cell nucleus involved in distinguishing the two genders.

**Sex determination**   The genetic or developmental means by which an individual develops as either a male or a female (or both).

**Sex-role reversal**   A situation in which "usual" male behaviors (e.g. in mammals) are displayed by females, and vice versa; also used in a technical sense to mean any situation in which sexual selection operates more intensely on females than on males.

**Sexual dichromatism**   A notable difference in color or color pattern between males and females of a given species.

**Sexual dimorphism**   A notable difference in phenotypic appearance (other than in the sex organs *per se*) between males and females of a given species.

**Sexual reproduction**   Reproduction involving the production and subsequent fusion of gametes.

**Sexual selection**   The differential ability of individuals of the two genders to acquire mates. Intrasexual selection refers to competition among members of the same sex over access to mates; intersexual selection refers to patterns of mate choice by males and females.

**Sibling species**   *See* **cryptic species**.

**Sister taxa**   Taxa stemming from the same node in a phylogenetic tree.

**Somatic cell**   Any cell in a multicellular organism other than those destined to become gametes.

**Species (biological)**   Groups of actually or potentially interbreeding individuals that are reproductively isolated from other such groups.

**Species tree**   A diagram of phylogenetic relationships among species (as to be distinguished from a gene tree, which for various reasons can differ somewhat in topology from the multi-locus composite or consensus species tree).

**Sperm**   A male gamete in animals.

**Squamate**   A member of a subset of reptiles that includes lizards and snakes.

**Stamen**   A male reproductive organ in a flower, typically arranged as a filament.

**Stigma**   The part of a flower's pistil that receives the pollen.

**Structural gene**   A gene that encodes a protein.

**Symbiont**   One of the participants in a symbiosis.

**Symbiosis**   Any close association between individuals of two or more species, sometimes but not necessarily restricted to collaborations that are mutually beneficial.

**Sympatric**   Inhabiting the same geographic area.

**Symplesiomorphy**   An ancestral character state shared by two or more descendent taxa.

**Synapomorphy**   A derived character state shared by two or more descendent taxa.

**Systematics**   The comparative study and classification of organisms, particularly with regard to their phylogenetic relationships.

**Systematist**   A scientist who practices systematics.

**Taxon** (pl. **taxa**).   A biotic lineage or entity deemed sufficiently distinct from other such lineages as to be worthy of a formal taxonomic name.

**Taxonomy**   The practice of naming and classifying organisms.

**Tertiary Period**   The geological time frame beginning about 65 million years ago and ending about two million years before the present.

**Tetrapod**   A vertebrate animal other than a fish.

**Thorax**   The region of the body between the head and the abdomen.

**Toxin**   A poisonous substance.

**Transfer RNA**    An RNA molecule that transfers an amino acid to a growing polypeptide chain during translation.

**Transposable element**    Any of a class of DNA sequences that can move from one chromosomal site to another, often replicatively.

**Tree of Life**    The complete phylogenetic history of life on the planet Earth.

**Ungulate**    Any large, hoof-bearing, grazing mammal.

**Unisexual**    Consisting of one sex only.

**Uterus**    A mammalian organ in which an embryo develops after implantation.

**Vaccine**    A suspension of dead or weakened microbes, or biochemical constituents thereof, injected into the body to immunize against that same pathogen.

**Venom**    A biologically produced poison or toxin.

**Vertebrate**    An animal that possesses a backbone.

**Vestigial**    Degenerate or rudimental; used of anatomical structures or functions that have become reduced in particular organisms during the evolutionary process.

**Vicariance**    The process by which a historical barrier to dispersal can lead to the evolutionary emergence of two or more closely related forms of animals or plants in different geographical areas.

**Virus**    A tiny, obligate intracellular parasite, incapable of autonomous replication, which utilizes the host cell's replicative machinery.

**Viviparous**    Producing live offspring by giving birth from within the body of a parent, a process known as viviparity or live-bearing.

**Warning coloration**    Conspicuous coloration that serves to advertise the noxious, unpalatable, or otherwise dangerous properties of an organism to a potential predator.

**W-chromosome**    In birds, the sex chromosome normally present in females only.

**Womb**    *See* **uterus**.

**X-chromosome**    The sex chromosome normally present as two copies in female mammals (the homogametic sex) but as only one copy in males (the heterogametic sex).

**Y-chromosome**    In mammals, the sex chromosome normally present in males only.

**Z-chromosome**    The sex chromosome normally present as two copies in male birds (the homogametic sex) but as only one copy in females (the heterogametic sex).

**Zygote**    Fertilized egg; the diploid cell arising from a union of male and female haploid gametes.

# REFERENCES AND FURTHER READING

## Chapter 1

Avise, J. C. 2002. *Genetics in the Wild*. Washington, D.C.: Smithsonian Institution Press. 2004. *Molecular Markers, Natural History, and Evolution* (2nd edn). Sunderland, MA: Sinaucr.

Baker, A. J. (ed.) 2000. *Molecular Methods in Ecology*. Oxford: Blackwell.

Darwin, C. 1859. *On the Origin of Species*. London: John Murray.

Dawkins, R. 2004. *The Ancestor's Tale: A Pilgrimage to the Dawn of Evolution*. New York: Houghton-Mifflin.

Dobzhansky, T. 1973. Nothing in biology makes sense except in the light of evolution. *Am. Biol. Teacher* **35**: 125–9.

Felsenstein, J. 2004. *Inferring Phylogenies*. Sunderland, MA: Sinauer.

Haeckel, E. 1866. *Generelle Morphologie der Organismen*. Berlin: Georg Reimer.

Hall, B. G. 2004. *Phylogenetic Trees Made Easy* (2nd edn). Sunderland, MA: Sinauer.

Hillis, D. M., C. Moritz, and B. K. Mable (eds) 1996. *Molecular Systematics* (2nd edn). Sunderland, MA: Sinauer.

Holder M. and P. O. Lewis 2003. Phylogeny estimation: traditional and Bayesian approaches. *Nature Genet*. **4**: 275–84.

Huelsenbeck, J. P. 2000. *MRBAYES: Bayesian Inferences of Phylogeny* [software]. Rochester, NY: University of Rochester.

Huelsenbeck, J. P. and B. Rannala 1997. Phylogenetic methods come of age: testing hypotheses in an evolutionary context. *Science* **276**: 227–232.

Li, W.-H. 1997. *Molecular Evolution*. Sunderland, MA: Sinauer.

Margoliash, E. 1963. Primary structure and evolution of cytochrome *c*. *Proc. Natl. Acad. Sci. USA* **50**: 672–9.

Nei, M. and S. Kumar 2000. *Molecular Evolution and Phylogenetics*. Oxford: Oxford University Press.

Rokas, A., B. L. Williams, N. King, and S. B. Carroll 2003. Genome-scale approaches to resolving incongruence in molecular phylogenies. *Nature* **425**: 798–804.

Simpson, G. G. 1945. The principles of classification and a classification of mammals. *Bull. Am. Mus. Nat. Hist*. **85**: 1–350.

Strimmer, K. and A. von Haeseler 1996. Quartet puzzling: A quartet maximum likelihood method for reconstructing tree topologies. *Molec. Biol. Evol.* **13**: 964–9.

Swofford, D. L. 2000. PAUP*: *Phylogenetic Analysis Using Parsimony and Other Methods* [software]. Sunderland, MA: Sinauer.

## Chapter 2

### Whence the toucan's bill?

Lanyon, S. M. and J. G. Hall 1994. Re-examination of barbet monophyly using mitochondrial-DNA sequence data. *Auk* **111**: 389–97.

Prum, R. O. 1988. Phylogenetic interrelationships of the barbets (Aves: Capitonidae) and toucans (Aves: Ramphastidae) based on morphology with comparisons to DNA-DNA hybridization. *Zool. J. Linn. Soc.* **92**: 313–43.

Sibley, C. G. and J. E. Ahlquist. 1986. Reconstructing bird phylogeny by comparing DNA's. *Scient. Am.* **254**(2): 82–3.

### The beak of the fish

Beer, G. R. de 1940. *Embryos and Ancestors*. Oxford: Clarendon Press.

Boughton, D. A., B. B. Collette, and A. R. McCune. 1991. Heterochrony in jaw morphology of needlefishes (Teleostei: Belonidae). *Syst. Zool.* **40**: 329–54.

Collette, B. B. and N. V. Parin. 1970. Needlefishes (Belonidae) of the Eastern Atlantic Ocean. *Atl. Rep.* **11**: 1–60.

Gould, S. J. 2000. *Ontogeny and Phylogeny*. Cambridge, MA: Harvard University Press.

Haeckel, E. 1866. *Generelle Morphologie der Organismen*. Berlin: Georg Reimer.

Lovejoy, N. R. 2000. Reinterpreting recapitulation: Systematics of needlefishes and their allies (Teleostei: Beloniformes). *Evolution* **54**: 1349–62.

### Snails' shell shapes

Collin, R. and R. Cipriani 2003. Dollo's law and the re-evolution of shell coiling. *Proc. R. Soc. Lond.* B**270**: 2551–5.

Dollo, L. 1893. Les lois de l'evolution. *Bull. Soc. Belge Géol. Pal. Hydr*. **7**: 164–6.

Gould, S. J. 1970. Dollo on Dollo's law: irreversibility and the status of evolutionary laws. *J. Hist. Biol.* **3**: 189–212.

Raff, R. A. 1996. *The Shape of Life: Genes, Development, and the Evolution of Animal Form*. Chicago: University of Chicago Press.

Vermeij, G. 1987. *Evolution and Escalation*. Princeton, NJ: Princeton University Press.

### More on snails' shell shapes

Asami, T., R. H. Cowie, and K. Ohbayashi 1998. Evolution of mirror images by sexually asymmetric mating behavior in hermaphroditic snails. *Am. Nat.* **152**: 225–36.

Gittenberger, E. 1988. Sympatric speciation in snails: A largely neglected model. *Evolution* **42**: 826–8.

Johnson, M. S., B. Clarke, and J. Murray 1990. The coil polymorphism in *Partula suturalis* does not favor sympatric speciation. *Evolution* **44**: 459–64.

Ueshima, R. and T. Asami 2003. Single-gene speciation by left-right reversal. *Nature* **425**: 679.

Vermeij, G. J. 1975. Evolution and distribution of left-handed and planispiral coiling in snails. *Nature* **254**: 419–20.

### Winged walkingsticks

Wagner, D. L. and J. K. Liebherr 1992. Flightlessness in insects. *Trends Ecol. Evol.* **7**: 216–20.

Whiting, M. F., S. Bradler, and T. Maxwell 2003. Loss and recovery of wings in stick insects. *Nature* **421**: 264–7.

### Hermits and kings

Cunningham, C. W., N. W. Blackstone, and L. W. Buss 1992. Evolution of king crabs from hermit crab ancestors. *Nature* **355**: 539–42.

Gould, S. J. 1992. We are all monkey's uncles. *Nat. Hist.* **101**(6): 14–21.

### True and false gharials

Brochu, C. A. 2001. Crocodylian snouts in space and time: Phylogenetic approaches toward adaptive radiation. *Am. Zool.* **41**: 564–85.

Gatesy, J. and G. D. Amato. 1992. Sequence similarity of 12S ribosomal segment of mitochondrial DNAs of gharial and false gharial. *Copeia* **1992**: 241–3.

Graybeal, A. 1994. Evaluating the phylogenetic utility of genes: A search for genes informative about deep divergences among vertebrates. *Syst. Biol.* **43**: 174–93.

Grigg, G. C., F. Seebacher, and C. E. Franklin (eds) 2001. *Crocodilian Biology and Evolution*. Chipping Norton, New South Wales, Australia: Surrey Beatty & Sons.

Harshman, J., C. J. Huddleston, J. P. Bollback, T. J. Parsons, and M. J. Braun 2003. True and false gharials: a nuclear gene phylogeny of Crocodylia. *Syst. Biol.* **52**: 386–402.

Hillis, D. M. 1987. Molecular versus morphological approaches to systematics. *A. Rev. Ecol. Syst.* **18**: 23–42.

Maddison, W. P. 1997. Gene trees in species trees. *Syst. Biol.* **46**: 523–36.

Norell, M. A. 1989. The higher level relationships of the extant Crocodylia. *J. Herpetol.* **23**: 325–35.

### Loss of limbs on the reptile tree

Caldwell, M. W. and M. S. Y. Lee 1997. A snake with legs from the marine Cretaceous of the Middle East. *Nature* **386**: 705–9.

Coates, M. and M. Ruta 2000. Nice snake, shame about the legs. *Trends Ecol. Evol.* **15**: 503–7.

Greer, A. E. 1991. Limb reduction in squamates: identification of the lineages and discussion of the trends. *J. Herpetol.* **25**: 166–73.

Kearney, M. and B. L. Stuart 2004. Repeated evolution of limblessness and digging heads in worm lizards revealed by DNA from old bones. *Proc. R. Soc. Lond.* B**271**: 1677–83.

Lande, R. 1978. Evolutionary mechanisms of limb loss in tetrapods. *Evolution* **32**: 73–92.

Pough, F. H. and 5 others 1998. *Herpetology*. Upper Saddle River, NJ: Prentice-Hall.

Vidal, N. and S. B. Hedges 2004. Molecular evidence for a terrestrial origin of snakes. *Proc. R. Soc. Lond.* B (suppl.)**271**: S226–9.

Walls, G. L. 1940. Ophthalmalogical implications for the early history of snakes. *Copeia* **1940**: 1–8.

Wiens, J. J. and J. L. Slingluff 2001. How lizards turn into snakes: a phylogenetic analysis of body-form evolution in anguid lizards. *Evolution* **55**: 2303–18.

### Fishy origins of tetrapods

Brinkmann, H., B. Venkatesh, S. Brenner, and A. Meyer 2004. Nuclear protein-coding genes support lungfish and not the coelacanth as the closest living relatives of land vertebrates. *Proc. Natl. Acad. Sci. USA* **101**: 4900–5.

Gorr, T., T. Kleinschmidt, and H. Fricke 1991. Close tetrapod relationship of the coelacanth *Latimeria* indicated by haemoglobin sequences. *Nature* **351**: 394–7.

Meyer, A. and A. C. Wilson 1990. Origin of tetrapods inferred from their mitochondrial DNA affiliation to lungfish. *J. Molec. Evol.* **31**: 359–64.

Sharp, P. M., A. T. Lloyd, and D. G. Higgins 1991. Coelacanth's relationships. *Nature* **353**: 218–19.

Stock, D. W., K. D. Moberg, L. R. Maxson, and G. S. Whitt 1991. A phylogenetic analysis of the 18S ribosomal RNA sequence of the coelacanth *Latimeria chalumnae*. *Env. Biol. Fishes* **32**: 99–117.

Takezaki, N., F. Figueroa, Z. Zaleska-Rutczynska, N. Takahata, and J. Klein 2004. The phylogenetic relationships of tetrapod, coelacanth, and lungfish revealed by the sequences of forty-four nuclear genes. *Molec. Biol. Evol.* **21**: 1512–24.

Thompson, K. S. 1991. *Living Fossil: The Story of the Coelacanth*. New York: Norton.

Zardoya, R., Y. Cao, M. Hasegawa, and A. Meyer 1998. Searching for the closest living relative(s) of tetrapods through evolutionary analyses of mitochondrial and nuclear data. *Molec. Biol. Evol.* **15**: 506–17.

### Panda ponderings

Flynn, J. J., M. A. Nedbal, J. W. Dragoo, and R. L. Honeycutt 2000. Whence the red panda? *Molec. Phylogen. Evol.* **17**: 190–9.

O.Brien, S. J. 1987. The ancestry of the giant panda. *Scient. Am.* **257**(5): 102–7.

O'Brien, S. J., W. G. Nash, D. E. Wildt, M. E. Bush, and R. E. Benveniste 1985. A molecular solution to the riddle of the giant panda's phylogeny. *Nature* **317**: 140–4.

Sarich, V. M. 1973. The giant panda is a bear. *Nature* **245**: 218–20.

Slattery, J. P. and S. J. O'Brien 1995. Molecular phylogeny of the red panda *(Ailurus fulgens)*. *J. Hered.* **86**: 413–22.

### Fossil DNA and extinct eagles

Brown, L. H. and D. Amadon 1968. *Eagles, Hawks and Falcons of the World*. London: Country Life.

Bunce, M. and 6 others. 2005. Ancient DNA provides new insights into the evolutionary history of New Zealand's extinct giant eagle. *PloS Biology* **3**: 44–6.

Hofreiter, M, D. Serre, H. N. Poinar, M. Kuch, and S. Pääbo 2001. Ancient DNA. *Nature Rev. Genet.* **2**: 353–9.

Nicholls, H. 2005. Ancient DNA comes of age. *PloS Biology* **3**: 192–6.

Worthy, T. H. and R. N. Holdaway 2002. *The Lost World of the Moa: Prehistoric Life of New Zealand*. Bloomington, IN: Indiana University Press.

### The Yeti's abominable phylogeny

Hergé, G. R. 1960. *Tintin in Tibet* [English version]. Belgium: Casterman.

Matthiessen, P. 1979. *The Snow Leopard*. London: Chatto & Windus.

Matthiessen, P. and T. Laird 1995. *East of Lo Monhong: In the Land of the Mustang*. Boston, MA: Shambala Publishers.

Milinkovitch, M. C., A. Caccone, and G. Amato 2004. Molecular phylogenetic analyses indicate extensive morphological convergence between the "yeti" and primates. *Molec. Phylogen. Evol.* **31**: 1–3.

## Chapter 3

### Light and dark mice

Dice, L. and P. M. Blossom 1937. Studies of mammalian ecology in southwestern North America, with special attention to the colors of desert mammals. *Publ. Carnegie Inst. Washington* **485**: 1–25.

Hoekstra, H. E. and M. W. Nachman 2003. Different genes underlie adaptive melanism in different populations of rock pocket mice. *Molec. Ecol.* **12**: 1185–94.

Nachman, M. W., H. E. Hoekstra, and S. L. D'Agostino 2003. The genetic basis of adaptive melanism in pocket mice. *Proc. Natl. Acad. Sci. USA* **100**: 5268–73.

### Sexual dichromatism

Andersson, M. 1994. *Sexual Selection*. Princeton, NJ: Princeton University Press.

Badyaev, A. V. and G. E. Hill 2003. Avian sexual dichromatism in relation to phylogeny and ecology. *A. Rev. Ecol. Evol. Syst.* **34**: 27–49.

Burns, K. J. 1998. A phylogenetic perspective on the evolution of sexual dichromatism in tanagers (Thraupidae): The role of female versus male plumage. *Evolution* **52**: 1219–24.

Kimball, R. T., E. L. Braun, J. D. Ligon, V. Lucchini, and E. Randi 2001. A molecular phylogeny of the peacock-pheasants (Galliformes: *Polyplectron* spp) indicates loss and reduction of ornamental traits and display behaviors. *Biol. J. Linn. Soc.* **73**: 187–98.

Kimball, R. T. and J. D. Ligon 1999. Evolution of avian plumage dichromatism from a proximate perspective. *Am. Nat.* **154**: 182–93.

Owens, I. P. F. and R. V. Short 1995. Hormonal basis of sexual dimorphism in birds: implications for new theories of sexual selection. *Trends Ecol. Evol.* **10**: 44–7.

Peterson, A. T. 1996. Geographic variation in sexual dichromatism in birds. *Bull. Br. Ornithol. Club* **116**: 156–72.

Price, T. and G. L. Birch 1996. Repeated evolution of sexual color dimorphism in passerine birds. *Auk* **133**: 842–8.

Wiens, J. 2001. Widespread loss of sexually selected traits: how the peacock lost its spots. *Trends Ecol. Evol.* **16**: 517–23.

### Dabbling into duck plumages

Delacour, J. and E. Mayr 1945. The family Anatidae. *Wilson Bull.* **57**: 2–55.

Omland, K. E. 1997. Examining two standard assumptions of ancestral reconstructions: repeated loss of dichromatism in dabbling ducks (Anatini). *Evolution* **51**: 1636–46.

Sibley, C. G. 1957. The evolutionary and taxonomic significance of sexual dimorphism and hybridization in birds. *Condor* **59**: 166–87.

### Specific avian color motifs

Allen, E. S. and K. E. Omland 2003. Novel intron phylogeny supports plumage convergence in orioles (*Icterus*). *Auk* **120**: 961–9.

Endler, J. A. and M. Théry 1996. Interacting effects of lek placement, display behavior, ambient light, and color patterns in three Neotropical forest-dwelling birds. *Am. Nat.* **148**: 421–52.

Hoekstra, H. E. and T. Price 2004. Parallel evolution is in the genes. *Science* **303**: 1779–81.

Mundy, N. I. and 5 others 2004. Conserved genetic basis of a quantitative plumage trait involved in mate choice. *Science* **303**: 1870–3.

Omland, K. E. and S. M. Lanyon 2000. Reconstructing plumage evolution in orioles (*Icterus*): Repeated convergence and reversal in patterns. *Evolution* **54**: 2119–33.

West-Eberhard, M. J. 2003. *Developmental Plasticity and Evolution*. New York: Oxford University Press.

The poisonous Pitohui

Diamond, J. 1994. Stinking birds and burning books. *Natural History* **103**(2): 4–12.

Dumbacher, J. P. and R. C. Fleischer 2001. Phylogenetic evidence for colour pattern convergence in toxic pitohuis: Müllerian mimicry in birds? *Proc. R. Soc. Lond.* B**268**: 1971–6.

Dumbacher, J. P. and S. Pruett-Jones 1996. Avian chemical defenses. *Curr. Ornithol.* **13**: 137–74.

Müller, F. 1879. *Ituna* and *Thyridia*: a remarkable case of mimicry in butterflies. *Trans. Entomol. Soc. Lond.* **1879**: xx–xxix.


Warning colorations in poison frogs

Daly, J. W. and 6 others 2002. Bioactive alkaloids of frog skin: combinatorial bioprospecting reveals that pumiliotoxins have an arthropod source. *Proc. Natl. Acad. Sci. USA* **99**: 13996–4001.

Myers, C. W. and J. W. Daly 1983. Dart-poison frogs. *Scient. Am.* **248**(2): 120–33.

Santos, J. C., L. A. Coloma, and D. C. Cannatella 2003. Multiple, recurring origins of aposematism and diet specialization in poison frogs. *Proc. Natl. Acad. Sci. USA* **100**: 12792–7.

Saporito, R. A. and 5 others 2004. Formicine ants: an arthropod source for the pumiliotoxin alkaloids of dendrobatid poison frogs. *Proc. Natl. Acad. Sci. USA* **101**: 8045–50.

Symula, R., R. Schulte, and K. Summers 2001. Molecular phylogenetic evidence for a mimetic radiation in Peruvian poison frogs supports a Müllerian mimicry hypothesis. *Proc. R. Soc. Lond.* B**268**: 2415–21.


Müllerian mimicry butterflies

Brower, A. V. Z. 1994. Rapid morphological radiation and convergence among races of the butterfly *Heliconius erato* inferred from patterns of mitochondrial DNA evolution. *Proc. Natl. Acad. Sci. USA* **91**: 6491–5.

    1996. Parallel race formation and the evolution of mimicry in *Heliconius* butterflies: a phylogenetic hypothesis from mitochondrial DNA sequences. *Evolution* **50**: 195–221.

Nijhout, H. F. 1991. *The Development and Evolution of Butterfly Wing Patterns*. Washington, DC: Smithsonian Institution Press.


Caterpillar colors and cryptic species

Frankie, G. W., A. Mata, and S. B. Vinson (eds) 2004. *Biodiversity Conservation in Costa Rica*. Berkeley, CA: University of California Press.

Hebert, P. D. N., A. Cywinska, S. L. Ball, and J. R. deWaard 2003. Biological identifications through DNA barcodes. *Proc. R. Soc. Lond.* B**270**: 313–21.

Hebert, P. D. N., E. H. Penton, J. M. Burns, D. H. Janzen, and W. Hallwachs 2004. Ten species in one: DNA barcoding reveals cryptic species in the neotropical skipper butterfly *Astraptes fulgerator. Proc. Natl. Acad. Sci. USA* **101**: 14812–17.

Tautz, D., P. Arctander, A. Minelli, R. H. Thomas, and A. P. Vogler 2003. A plea for DNA taxonomy. *Trends Ecol. Evol*. **18**: 70–4.

Wilson, E. O. 1992. *The Diversity of Life*. New York: Norton.

## Chapter 4

### The chicken or the egg?

Meyer, A. and R. Zardoya 2003. Recent advances in the (molecular) phylogeny of vertebrates. *A. Rev. Ecol. Evol. Syst*. **34**: 311–38.

Gill, F. B. 1990. *Ornithology* (2nd edn). New York: W. H. Freeman & Co.

### The avian nest

Bennett, P. M. and I. P. F. Owens 2002. *Evolutionary Ecology of Birds*. Oxford: Oxford University Press.

Owens, I. P. F. and P. M. Bennett 1995. Ancient ecological diversification explains life-history variation among living birds. *Proc. R. Soc. Lond*. B**261**: 227–32.

Sheldon, F. H., L. A. Whittingham, and D. W. Winkler 1999. A comparison of cytochrome *b* and DNA hybridization data bearing on the phylogeny of swallows (Aves: Hirundinidae). *Molec. Phylogen. Evol*. **11**: 320–31.

Winkler, D. W. and F. H. Sheldon 1993. Evolution of nest construction in swallows (Hirundinidae): A molecular phylogenetic perspective. *Proc. Natl. Acad. Sci. USA* **90**: 5705–7.

### Egg dumping and foster parentage

Aragon, S., A. P. Møller, J. J. Soler and M. Soler 1999. Molecular phylogeny of cuckoos supports a polyphyletic origin of brood parasitism. *J. Evol. Biol*. **12**: 495–506.

Lanyon, S. M. 1992. Interspecific brood parasitism in blackbirds (Icterinae): A phylogenetic perspective. *Science* **255**: 77–9.

Sorenson, M. D., K. M. Sefc, and R. B. Payne 2003. Speciation by host switch in brood parasitic indigobirds. *Nature* **424**: 928–31.

### Egg laying and live bearing

Blackburn, D. G. 1992. Convergent evolution of viviparity, matrotrophy and specializations for fetal nutrition in reptiles and other vertebrates. *Am. Zool*. **32**: 313–21.

Bull, J. J. and E. L. Charnov 1985. On irreversible evolution. *Evolution* **39**: 1149–55.

Dulvy, N. K. and J. D. Reynolds 1997. Evolutionary transitions among egg-laying, live-bearing and maternal inputs in sharks and rays. *Proc. R. Soc. Lond.* B**264**: 1309–15.

Lee, M. S. and R. Shine 1998. Reptilian viviparity and Dollo's law. *Evolution* **52**: 1441–50.

Neill, W. T. 1964. Viviparity in snakes: some ecological and zoogeographical considerations. *Am. Nat.* **98**: 35–55.

Rouse, G. and K. Fitzhugh 1994. Broadcasting fables: is external fertilization really primitive? *Zool. Scr.* **23**: 271–312.

Surget-Groba, Y. and 13 others 2001. Intraspecific phylogeography of *Lacerta vivipara* and the evolution of viviparity. *Molec. Phylogen. Evol.* **18**: 449–59.

### Piscine placentas

Darwin, C. 1859. *On the Origin of Species.* London: John Murray.

Mateos, M., O. I. Sanjur, and R. C. Vrijenhoek 2002. Historical biogeography of the livebearing fish genus *Poeciliopsis* (Poeciliidae: Cyprinodontiformes). *Evolution* **56**: 972–84.

Nilsson, D.-E. and S. Pelger. 1994. A pessimistic estimate of the time required for an eye to evolve. *Proc. R. Soc. Lond.* B**256**: 53–8.

Reznick, D. N., M. Mateos, and M. S. Springer 2002. Independent origins and rapid evolution of the placenta in the fish genus *Poeciliopsis. Science* **298**: 1018–20.

Rossant, J. and J. C. Cross. 2001. Placental development: lessons from mouse mutants. *Nature Rev. Genet.* **2**: 538–48.

### Male pregnancy

Jones, A. G. and J. C. Avise 2001. Mating systems and sexual selection in male-pregnant pipefishes and seahorses: insights from microsatellite-based studies of maternity. *J. Heredity* **92**: 150–8.

Lourie, S. A., A. Vincent, and H. J. Hall 1999. *Seahorses: An Identification Guide to the World's Species and Their Conservation.* London: Project Seahorse.

Vincent, A., I. Ahnesjö, A. Berglund, and G. Rosenqvist 1992. Pipefishes and seahorses: are they all sex role reversed? *Trends Ecol. Evol.* **7**: 237–41.

Wilson, A. B., I. Ahnesjö, A. Vincent, and A. Meyer 2003. The dynamics of male brooding, mating patterns, and sex roles in pipefishes and seahorses (family Syngnathidae). *Evolution* **57**: 1374–86.

Wilson, A. B., A. Vincent, I. Ahnesjö, and A. Meyer 2001. Male pregnancy in seahorses and pipefishes (family Syngnathidae): rapid diversification of paternal brood pouch morphology inferred from a molecular phylogeny. *J. Heredity* **92**: 159–66.

### Living and reproducing by the sword

Basolo, A. L. 1990. Female preference predates the evolution of the sword in swordtail fish. *Science* **250**: 808–10.

1995. Phylogenetic evidence for the role of pre-existing bias in sexual selection. *Proc. R. Soc. Lond.* B**259**: 307–11.

Basolo, A. L. and G. Alcaraz 2003. The turn of the sword: length increases male swimming costs in swordtails. *Proc. R. Soc. Lond.* B**270**: 1631–6.

Endler, J. A. and A. L. Basolo 1998. Sensory ecology, receiver biases and sexual selection. *Trends Ecol. Evol.* **13**: 415–20.

Meyer, A., J. M. Morrissey, and M. Schartl 1994. Recurrent origin of a sexually selected trait in *Xiphophorus* fishes inferred from a molecular phylogeny. *Nature* **368**: 539–42.

Schluter, D., T. Price, A. Mooers, and D. Ludwig 1997. Likelihood of ancestor states in adaptive evolution. *Evolution* **51**: 1699–711.

### Brood care in Jamaican land crabs

Burggren, W. W. and B. R. McMahon (eds) 1988. *Biology of the Land Crabs.* Cambridge: Cambridge University Press.

Hedges, S. B. 1996. Historical biogeography of West Indian vertebrates. *A. Rev. Ecol. Syst.* **27**: 163–96.

Schubart, C. D., R. Diesel, and S. B. Hedges 1998. Rapid evolution to terrestrial life in Jamaican crabs. *Nature* **393**: 363–5.

### Social parasitism of butterflies on ants

Als, T. D. and 8 others 2004. The evolution of alternative parasitic life histories in large blue butterflies. *Nature* **432**: 386–90.

Hölldobler, B. and E. O. Wilson. 1990. *The Ants.* Berlin: Springer.

Pullin, A.S. (ed.) 1995. *Ecology and Conservation of Butterflies.* London: Chapman & Hall.

Thomas, J. A. and J. Settele 2004. Butterfly mimics of ants. *Nature* **432**: 283–4.

### Parthenogenetic lizards, geckos, and snakes

Avise, J. C., J. M. Quattro, and R. C. Vrijenhoek 1992. Molecular clones within organismal clones. *Evol. Biol.* **26**: 225–46.

Dawley, R. M. and J. P. Bogart (eds) 1989. *Evolution and Ecology of Unisexual Vertebrates.* Albany, NY: New York State Museum.

Densmore, L. D. III, C. C. Moritz, J. W. Wright, and W. M. Brown 1989. Mitochondrial-DNA analyses and the origin and relative age of parthenogenetic lizards (genus *Cnemidophorus*). IV. Nine *sexlineatus*-group unisexuals. *Evolution* **43**: 969–83.

Dessauer, H. C. and C. J. Cole 1989. Diversity between and within nominal forms of unisexual teiid lizards. In: *Evolution and Ecology of Unisexual Vertebrates*, R. M. Dawley and J. P. Bogart (eds), pp. 49–71. Albany, NY: New York State Museum.

Moritz, C. C. 1991. The origin and evolution of parthenogenesis in *Heteronotia binoei* (Gekkonidae): Evidence for recent and localized origins of widespread clones. *Genetics* **129**: 211–19.

Moritz, C. C. and 9 others 1989. Genetic diversity and the dynamics of hybrid partheno-genesis in *Cnemidophorus* (Teiidae) and *Heteronotia* (Gekkonidae). In: *Evolution and Ecology of Unisexual Vertebrates*, R. M. Dawley and J. P. Bogart (eds), pp. 87–112. Albany, NY: New York State Museum.

Quattro, J. M., J. C. Avise, and R. J. Vrijenhoek 1992. An ancient clonal lineage in the fish genus *Poeciliopsis* (Atheriniformes: Poeciliidae). *Proc. Natl. Acad. Sci. USA* **89**: 348–52.

## Of monkeyflowers and hummingbirds

Beardsley, P. M., A. Yen, and R. G. Olmstead 2003. AFLP phylogeny of *Mimulus* section *Erythranthe* and the evolution of hummingbird pollination. *Evolution* **57**: 1397–410.

Grant, K.A. and V. Grant, 1968. *Hummingbirds and Their Flowers*. New York: Columbia University Press.

Schemske, D. W. and H. D. Bradshaw, Jr. 1999. Pollinator preference and the evolution of floral traits in monkeyflowers (*Mimulus*). *Proc. Natl. Acad. Sci. USA* **96**: 11910–15.

Stebbins, G. L. 1970. Adaptive radiation of reproductive characteristics in Angiosperms. I. Pollination mechanisms. *A. Rev. Ecol. Syst.* **1**: 307–26.

Weller, S. G. and A. K. Sakai 1999. Using phylogenetic approaches for the analysis of plant breeding system evolution. *A. Rev. Ecol. Syst.* **30**: 167–99.

## Delayed implantation

Bininda-Emonds, O. R. P., J. L. Gittleman, and A. Purvis 1999. Building large trees by combining phylogenetic information: a complete phylogeny of the extant Carnivora (Mammalia). *Biol. Rev. Camb. Philos. Soc.* **74**: 143–75.

Birkhead, T. R. and A. P. Møller 1993. Sexual selection and the temporal separation of reproductive events: sperm storage data from reptiles, birds and mammals. *Biol. J. Linn. Soc.* **50**: 295–311.

Lindenfors, P., L. Dalen, and A. Angerbjörn. 2003. The monophyletic origin of delayed implantation in carnivores and its implications. *Evolution* **57**: 1952–6.

Mead, R. A. 1989. The physiology and evolution of delayed implantation in carnivores. In: *Carnivore Behavior, Ecology, and Evolution*, J. L. Gittleman (ed.), pp. 437–64. Ithaca, NY: Cornell University Press.

Renfree, M. B. 1978. Embryonic diapause in mammals: a developmental strategy. In: *Dormancy and Developmental Arrest*, M. E. Clutter (ed.), pp. 1–46. New York: Academic Press.

Thom, M. D., D. D. P. Johnson, and D. W. Macdonald 2004. The evolution and maintenance of delayed implantation in the Mustelidae (Mammalia: Carnivora). *Evolution* **58**: 175–83.

## Chapter 5

### The kangaroo's bipedal hop

Burk, A., M. Westerman, and M. Springer 1998. The phylogenetic position of the musky rat-kangaroo and the evolution of bipedal hopping in kangaroos (Macropodidae: Diprotodontia). *Syst. Biol.* **47**: 457–74.

Marshall, L. G. 1974. Why kangaroos hop. *Nature* **248**: 174–6.

Szalay, F. S. 1994. *The Evolutionary History of Marsupials and an Analysis of Osteological Characters.* Cambridge: Cambridge University Press.

### Powered flight in winged mammals

Adkins, R. M. and R. L. Honeycutt 1991. Molecular phylogeny of the superorder Archonta. *Proc. Nat. Acad. Sci. USA* **88**: 10317–21.

Bailey, W. J., J. L. Slighton, and M. Goodman 1992. Rejection of the "flying primate" hypothesis by phylogenetic evidence from the ε-globin gene. *Science* **256**: 86–9.

Baker, R. J., M. J. Novacek, and N. B. Simmons 1991. On the monophyly of bats. *Syst. Zool.* **40**: 216–31.

Mindell, D. P., C. W. Dick, and R. J. Baker 1991. Phylogenetic relationships among megabats, microbats, and primates. *Proc. Natl. Acad. Sci. USA* **88**: 10322–6.

Pettigrew, J. D. 1986. Flying primates? Megabats have the advanced pathway from eye to midbrain. *Science* **231**: 1304–6.

Teeling, E. C. and 5 others 2000. Molecular evidence regarding the origin of echolocation and flight in bats. *Nature* **403**: 188–92.

2005. A molecular phylogeny for bats illuminates biogeography and the fossil record. *Science* **307**: 580–4.

Van Den Bussche, R. A., R. J. Baker, J. P. Huelsenbeck, and D. M. Hillis 1998. Base compositional bias and phylogenetic analyses: A test of the "flying DNA" hypothesis. *Molec. Phylogen. Evol.* **13**: 408–16.

### Magnetotaxis in bacteria

DeLong, E. F., R. B. Frankel, and D. A. Bazylinski 1993. Multiple evolutionary origins of magnetotaxis in bacteria. *Science* **259**: 803–6.

Frankel, R. B. and R. P. Blakemore (eds) 1990. *Iron Biominerals.* New York: Plenum Press.

Stackebrandt, E. and M. Goodfellow (eds) 1991. *Nucleic Acid Techniques in Bacterial Systematics.* New York: Wiley.

### Cetacean origins

Graur, D. and D. C. Higgins 1994. Molecular evidence for the inclusion of Cetaceans within the order Artiodactyla. *Molec. Biol. Evol.* **11**: 357–64.

Milinkovitch, M. C. and J. G. M. Thewissen 1997. Even-toed fingerprints on whale ancestry. *Nature* **388**: 622–3.

Montgelard, C., F. M. Catzeflis, and E. Douzery 1997. Phylogenetic relationships of artiodactyls and cetaceans as deduced from the comparison of cytochrome *b* and 12S rRNA mitochondrial sequences. *Molec. Biol. Evol.* **14**: 550–9.

Nikaido, M., A. P. Rooney, and N. Okada 1999. Phylogenetic relationships among certartiodactyls based on insertions of short and long interspersed elements: Hippopotamuses are the closest extant relatives of whales. *Proc. Natl. Acad. Sci. USA* **96**: 10261–6.

O'Leary, M. A. 2001. The phylogenetic position of cetaceans: further combined data analyses, comparisons with the stratigraphic record and a discussion of character optimization. *Am. Zool.* **41**: 487–506.

Shimamura, M. and 8 others 1997. Molecular evidence from retroposons that whales form a clade within even-toed ungulates. *Nature* **388**: 666–70.

Ursing, B. W. and U. Arnason 1998. Analyses of mitochondrial genomes strongly support a hippopotamus-whale clade. *Proc. R. Soc. Lond.* B**265**: 2251–5.

### Feeding and echolocation in whales

Hasegawa, M., J. Adachi, and M. C. Milinkovitch 1997. Novel phylogeny of whales supported by total molecular evidence. *J. Molec. Evol.* **44**: S117–20.

Milinkovitch, M. C. 1995. Molecular phylogeny of cetaceans prompts revision of morphological transformations. *Trends Ecol. Evol.* **10**: 328–34.

Nikaido, M. and 10 others 2001. Retroposon analysis of major cetacean lineages: The monophyly of toothed whales and the paraphyly of river dolphins. *Proc. Natl. Acad. Sci. USA* **98**: 7384–9.

### The phylogeny of thrush migration

Berthold, P. 2003. *Avian Migration*. New York: Springer.

Outlaw, D. C., G. Voelker, B. Mila, and D. J. Girman 2003. Evolution of long-distance migration in and historical biogeography of *Catharus* thrushes: a molecular phylogenetic approach. *Auk* **120**: 299–310.

### Pufferfish inflation

Wainwright, P. C. and R. G. Turingan 1997. Evolution of pufferfish inflation behavior. *Evolution* **51**: 506–18.

Winterbottom, R. 1974. The familial phylogeny of the Tetraodontiformes (Acanthopterygii: Pisces) as evidenced by their comparative myology. *Smithsonian Contrib. Zool.* **155**: 1–201.

Eusociality in shrimp

Danforth, B. N., L. Conway, and S. Ji 2003. Phylogeny of eusocial *Lasioglossum* reveals multiple losses of eusociality within a primitively eusocial clade of bees (Hymenoptera: Halictidae). *Syst. Biol.* **52**: 23–36.

Duffy, J. E. 1996. Eusociality in a coral-reef shrimp. *Nature* **381**: 512–4.

Duffy, J. E., C. L. Morrison, and R. Ríos 2000. Multiple origins of eusociality among sponge-dwelling shrimps (*Synalpheus*). *Evolution* **54**: 503–16.

Hamilton, W. D. 1964. The genetical evolution of social behavior I, II. *J. Theor. Biol.* **7**: 1–52.

Queller, D. C. and J. E. Strassmann 1998. Kin selection and social insects. *BioScience* **48**: 165–75.

Sherman, P. W., J. U. M. Jarvis, and R. D. Alexander (eds) 1991. *The Biology of the Naked Mole-Rat*. Princeton, NJ: Princeton University Press.

Wilson, E. O. 1975. *Sociobiology*. Cambridge, MA: Belknap Press.


Evolutionary reversals of salamander lifecycles

Chippindale, P. T., R. M. Bonett, A. S. Baldwin, and J. J. Wiens 2004. Phylogenetic evidence for a major reversal of life-history evolution in plethodontid salamanders. *Evolution* **58**: 2809–22.

Duellman, W. E. and L. Trueb 1986. *Biology of Amphibians*. New York: McGraw-Hill.

Hall, B. K. and M. H. Wake (eds) 1999. *The Origin and Evolution of Larval Forms*. San Diego, CA: Academic Press.

Mueller, R. L., J. R. Macey, M. Jaekel, D. B. Wake, and J. L. Boore 2004. Morphological homoplasy, life history evolution, and historical biogeography of plethodontid salamanders inferred from complete mitochondrial genomes. *Proc. Natl. Acad. Sci. USA* **101**: 13820–5.

Porter, M. L. and K. A. Crandall 2003. Lost along the way: the significance of evolution in reverse. *Trends Ecol. Evol.* **18**: 541–7.

Pough, F. H., C. M. Janis, and J. B. Heiser 2001. *Vertebrate Life*. Upper Saddle River, NJ: Prentice Hall.

Titus, T. A. and A. Larson 1996. Molecular phylogenetics of desmognathine salamanders (Caudata: Plethodontidae): a reevaluation of evolution in ecology, life history, and morphology. *Syst. Biol.* **45**: 451–71.


Dichotomous life histories in marine invertebrates

Collin, R. 2004. Phylogenetic effects, the loss of complex characters, and the evolution of development in calyptraeid gastropods. *Evolution* **58**: 1488–502.

Hart, M. W., M. Byrne, and M. J. Smith 1997. Molecular phylogenetic analysis of life-history evolution in asterinid starfish. *Evolution* **5**: 1848–61.

McHugh, D. and G. W. Rouse 1998. Life history evolution of marine invertebrates: new views from phylogenetic systematics. *Trends Ecol. Evol.* **13**: 182–6.

Reid, D. G. 1990. A cladistic phylogeny of the genus *Littorina* (Gastropoda): implications for evolution of reproductive strategies and for classification. *Hydrobiologia* **193**: 1–19.

Schulze, S. R., S. A. Rice, J. L. Simon, and S. A. Karl 2000. Evolution of poecilogony and the biogeography of North American populations of the polychaete *Streblospio*. *Evolution* **54**: 1247–59.

Strathmann, R. R. 1985. Feeding and nonfeeding larval development and life-history in marine invertebrates. *A. Rev. Ecol. Syst.* **16**: 339–61.

Villinski, J. T., J. C. Villinski, M. Byrne, and R. A. Raff 2002. Convergent maternal provisioning and life-history evolution in echinoderms. *Evolution* **56**: 1764–75.

### Adaptive radiations in island lizards

Losos, J. B., T. R. Jackman, A. Larson, K. de Queiroz, and L. Rodríguez-Schettino 1998. Contingency and determinism in replicated adaptive radiations of island lizards. *Science* **279**: 2115–18.

Losos, J. B. and 8 others 2003. Niche lability in the evolution of a Caribbean lizard community. *Nature* **423**: 542–5.

Miles, D. B. and A. E. Dunham 1996. The paradox of the phylogeny: character displacement of analyses of body size in island *Anolis*. *Evolution* **50**: 594–603.

Roughgarden, J. 1995. *Anolis Lizards of the Caribbean. Ecology, Evolution, and Plate Tectonics.* Oxford: Oxford University Press.

Schoener, T. W. 1969. Size patterns in West Indian *Anolis* lizards: I. Size and species diversity. *Syst. Zool.* **18**: 386–401.

### Spiders' web-building behaviors

Blackledge, T. A. and R. G. Cillespie 2004. Convergent evolution of behavior in an adaptive radiation of Hawaiian web-building spiders. *Proc. Natl. Acad. Sci. USA* **101**: 16228–33.

Schluter, D. 2000. *The Ecology of Adaptive Radiation.* New York: Oxford University Press.

Shear, W. A. 1986. *Spiders: Webs, Behavior, and Evolution.* Palo Alto, CA: Stanford University Press.

Wagner, W. L. and V. A. Funk (eds) 1995. *Hawaiian Biogeography: Evolution on a Hot Spot Archipelago.* Washington, DC: Smithsonian Institution Press.

### Lichen lifestyles

Ahmadjian, V. 1967. *The Lichen Symbiosis.* Waltham, MA: Blaisdell.

Gargas, A., P. T. DePriest, M. Grube, and A. Tehler 1995. Multiple origins of lichen symbioses in fungi suggested by SSU rRNA phylogeny. *Science* **268**: 1492–5.

Goff, J. (ed.) 1983. *Algal Symbiosis.* Cambridge: Cambridge University Press.

### Chapter 6

Foregut fermentation

Grajal, A., S. D. Strahl, R. Parra, M. G. Dominguez, and A. Neher 1989. Foregut fermentation in the hoatzin, a neotropical leaf-eating bird. *Science* **245**: 1236–8.

Irwin, D. M., E. M. Prager, and A. C. Wilson 1992. Evolutionary genetics of ruminant lysozymes. *Anim. Genet.* **23**: 193–202.

Kornegay, J. R., J. W. Schilling, and A. C. Wilson 1994. Molecular adaptation of a leaf-eating bird: Stomach lysozyme of the hoatzin. *Molec. Biol. Evol.* **11**: 921–8.

Stewart, C.-B., J. W. Schilling, and A. C. Wilson 1987. Adaptive evolution in the stomach lysozymes of foregut fermenters. *Nature* **330**: 401–4.

Swanson, K. W., D. M. Irwin, and A. C. Wilson 1991. Stomach lysozyme gene of the langur monkey: tests for convergence and positive selection. *J. Molec. Evol.* **33**: 418–25.

Snake venoms

Fry, B. G. and W. Wüster 2004. Assembling an arsenal: origin and evolution of the snake venom proteome inferred from phylogenetic analysis of toxin sequences. *Molec. Biol. Evol.* **21**: 870–83.

Greene, H. W. 1997. *Snakes: The Evolution of Mystery in Nature*. Berkeley, CA: University of California Press.

Jackson, K. 2003. The evolution of venom-delivery systems in snakes. *Zool. J. Linn. Soc.* **137**: 337–54.

Kelly, C. M. R., N. P. Barker, and M. H. Willet 2003. Phylogenetics of advanced snakes (Caenophidia) based on four mitochondrial genes. *Syst. Biol.* **52**: 439–59.

Slowinski, J. B. and R. Lawson 2002. Snake phylogeny: evidence from nuclear and mitochondrial genes. *Molec. Phylogen. Evol.* **24**: 194–202.

Underwood, G. 1997. An overview of venomous snake evolution. In: *Venomous Snakes: Ecology, Evolution and Snakebite*, R. S. Thorpe, W. Wüster, and A. Malhotra (eds), pp. 1–13. [Symposium of the Zoological Society of London, No. 70.] Oxford: Clarendon Press.

Vidal, N. 2002. Colubroid systematics: evidence for an early appearance of the venom apparatus followed by extensive evolutionary tinkering. *J. Toxicol. Toxin Rev.* **21**: 21–41.

Antifreeze proteins in anti-tropical fishes

Bargelloni, L., S. Marcato, L. Zane, and T. Patarnello 2000. Mitochondrial phylogeny of notothenioids: a molecular approach to Antarctic fish evolution and biogeography. *Syst. Biol.* **49**: 114–29.

Bargelloni, L. and 5 others 1994. Molecular evolution at subzero temperatures: mitochondrial and nuclear phylogenies of fishes from Antarctica (Suborder Notothenioidei), and the evolution of antifreeze glycopeptides. *Molec. Biol. Evol.* **11**: 854–63.

Chen, L., A. L. DeVries, and C.-H. C. Cheng 1997. Convergent evolution of antifreeze glycoproteins in Antarctic notothenioid fish and Arctic cod. *Proc. Natl. Acad. Sci. USA*
    **94**: 3817–22.

### Warm-bloodedness in fishes

Block, B. A. and R. J. Finnerty 1994. Endothermy in fishes: A phylogenetic analysis of
    constraints, predispositions, and selection pressures. *Environ. Biol. Fish.* **40**: 283–302.
Block, B. A., R. J. Finnerty, A. F. R. Stewart, and J. Kidd 1993. Evolution of endothermy in
    fish: Mapping physiological traits on a molecular phylogeny. *Science* **260**: 210–14.
Bennett, A. F. and J. A. Ruben. 1979. Endothermy and activity in vertebrates. *Science* **206**:
    649–54.
Carey, F. G., J. M. Teal, J. W. Kanwisher, and K. D. Lawson 1971. Warm-bodied fish. *Am.
    Zool.* **11**: 137–45.

### Electrical currents

Alves-Gomes, J. A., G. Orti, M. Haygood, W. Heiligenberg, and A. Meyer 1995. Phylogenetic
    analysis of the South American electric fishes (order Gymnotiformes) and the evolution of their electrogenic system: a synthesis based on morphology, electrophysiology,
    and mitochondrial sequence data. *Molec. Biol. Evol.* **12**: 298–318.
Helfman, G. S., B. B. Collette, and D. E. Facey 1997. *The Diversity of Fishes*. Malden, MA:
    Blackwell.
Hopkins, C. D., N. C. Comfort, J. Bastian, and A. H. Bass 1990. A functional analysis of sexual
    dimorphism in an electric fish, *Hypopomus pinnicaudatus*, order Gymnotiformes.
    *Brain Behav. Evol.* **35**: 350–67.
Lavoué, S., J. P. Sullivan, and C. D. Hopkins 2003. Phylogenetic utility of the first two introns
    of the S7 ribosomal protein gene in African electric fishes (Mormyroidea: Teleostei)
    and congruence with other molecular markers. *Biol. J. Linn. Soc.* **78**: 273–92.
Moller, P. 1995. *Electric Fishes: History and Behavior*. London: Chapman & Hall.
Sullivan, J. P., S. Lavoué, M. E. Arnegard, and C. D. Hopkins 2004. AFLPs resolve phylogeny
    and reveal mitochondrial introgression within a species flock of African electric fish
    (Mormyroidea: Teleostei). *Evolution* **58**: 825–41.
Sullivan, J. P., S. Lavoué, and C. D. Hopkins 2000. Molecular systematics of the African
    electric fishes (Mormyroidea: Teleostei) and a model for the evolution of their electric
    organs. *J. Exp. Biol.* **203**: 665–83.

### The Xs and Ys of sex determination

Bull, J. J. 1983. *Evolution of Sex Determining Mechanisms*. Menlo Park, CA: Benjamin
    Cummings.
Charlesworth, B. 1991. The evolution of sex chromosomes. *Science* **251**: 1030–3.

Ghiselin, M. T. 1969. The evolution of hermaphroditism among animals. *Q. Rev. Biol.* **44**: 189–208.

Graves, J. A. M. and S. Shetty 2001. Sex from W to Z: Evolution of vertebrate sex chromosomes and sex determining factors. *J. Exp. Zool.* **290**: 449–62.

Mank, J. E., D. E. L. Promislow, and J. C. Avise 2005. Evolution of sex-determining mechanisms in teleost fishes. *Biol. J. Linn. Soc., in press.*

Miya, M. and 11 others 2003. Major patterns of higher teleostean phylogenies: a new perspective based on 100 complete mitochondrial DNA sequences. *Molec. Phylogen. Evol.* **26**: 121–38.

Ohno, S. 1967. *Sex Chromosomes and Sex-linked Genes.* New York: Springer-Verlag.

Saitoh, K., M. Miya, J. G. Inoue, N. B. Ishiguro, and M. Nishida 2003. Mitochondrial genomics of Ostariophysan fishes: perspectives on phylogeny and biogeography. *J. Molec. Evol.* **56**: 464–72.

Solari, A. J. 1994. *Sex Chromosomes and Sex Determination in Vertebrates.* Boca Raton, FL: CRC Press.

The eyes have it

Darwin, C. 1859. *On the Origin of Species.* London: John Murray.

Gehring, W. J. 2000. Reply to Meyer-Rochow. *Trends Genet.* **16**: 245.

2005. New perspectives on eye development and the evolution of eyes and photoreceptors. *J. Heredity* **96**: 171–84.

Gehring, W. J. and K. Ikeo 1999. *Pax6*: Mastering eye morphogenesis and eye evolution. *Trends Genet.* **15**: 371–7.

Halder, G., P. Callaerts, and W. J. Gehring 1995. Induction of ectopic eyes by targeted expression of the eyeless gene in *Drosophila. Science* **267**: 1788–92.

Salvini-Plawen, L. and E. Mayr 1961. On the evolution of photoreceptors and eyes. In: *Evolutionary Biology*, M. K. Hecht, W. C. Steere, and B. Wallace (eds), pp. 207–63. New York: Plenum Press.

Two types of bodies

Brusca, R. C. and G. J. Brusca 2003. *Invertebrates.* Sunderland, MA: Sinauer.

Finnerty, J. R., K. Pang, P. Burton, D. Paulson, and M. Q. Martindale 2004. Origins of bilateral symmetry: *Hox* and *Dpp* expression in a sea anemone. *Science* **304**: 1335–7.

Hadzi, J. 1963. *The Evolution of the Metazoa.* Oxford: Pergamon Press.

Nielsen, C. 2001. *Animal Evolution: Interrelationships of the Living Phyla.* Oxford: Oxford University Press.

Tudge, C. 2000. *The Variety of Life.* Oxford: Oxford University Press.

Willmer, P. 1990. *Invertebrate Relationships: Patterns in Animal Evolution.* Cambridge: Cambridge University Press.

The phylogenomics of DNA repair

Bernstein, C. and H. Bernstein 1991. *Aging, Sex, and DNA Repair*. New York: Academic Press.

Eisen, J. A. and P. C. Hanawalt 1999. A phylogenomic study of DNA repair genes, proteins, and processes. *Mutation Res.* **435**: 171–213.

Hanawalt, P. C., P. K. Cooper, A. K. Ganesan, and C. A. Smith 1979. DNA repair in bacteria and mammalian cells. *A. Rev. Biochem.* **48**: 783–836.

Lander, E. S. and 243 others. 2001. Initial sequencing and analysis of the human genome. *Nature* **409**: 860–921.

Venter, J. C. and 273 others. 2001. The sequence of the human genome. *Science* **291**: 1304–53.

Roving nucleic acids

Arnold, M. L. 1997. *Natural Hybridization and Evolution*. New York: Oxford University Press.

Bushman, F. 2002. *Lateral DNA Transfer*. Cold Spring Harbor, NY: Cold Spring Harbor Laboratory Press.

Herédia, F., E. L. S. Loreto, and V. L. S. Valente 2004. Complex evolution of gypsy in drosophilid species. *Molec. Biol. Evol.* **21**: 1831–42.

Margulis, L. 1995. *Symbiosis in Cell Evolution: Microbial Communities in the Archaean and Proterozoic Eons* (2nd edn). San Francisco: W. H. Freeman & Co.

Raymond, J., O. Zhaxybayeva, J. P. Gogarten, S. Y. Gerdes, and R. E. Blankenship 2002. Whole-genome analysis of photosynthetic prokaryotes. *Science* **298**: 1616–20.

Rivera, M. C. and J. A. Lake 2004. The ring of life provides evidence for a genome fusion origin of eukaryotes. *Nature* **431**: 152–5.

Woese, C. R. and G. E. Fox 1977. Phylogenetic structure of the prokaryotic domain: the primary kingdoms. *Proc. Natl. Acad. Sci. USA* **74**: 5088–90.

Host-to-parasite gene transfer

Barkman, T. J., S.-H. Lim, K. M. Salleh, and J. Nais 2004. Mitochondrial DNA sequences reveal the photosynthetic relatives of *Rafflesia*, the world's largest flower. *Proc. Natl. Acad. Sci. USA* **101**: 787–92.

Bergthorsson, U., K. L. Adams, B. Thomason, and J. D. Palmer 2003. Widespread horizontal transfer of mitochondrial genes in flowering plants. *Nature* **424**: 197–201.

Bergthorsson, U., A. O. Richardson, G. J. Young, L. R. Goertzen, and J. D. Palmer 2004. Massive horizontal transfer of mitochondrial genes from diverse land plant donors to the basal angiosperm *Amborella*. *Proc. Natl. Acad. Sci. USA* **101**: 17747–52.

Davis, C. C. and K. J. Wurdack 2004. Host-to-parasite gene transfer in flowering plants: phylogenetic evidence from Malpighiales. *Science* **305**: 676–8.

Kuijt, J. 1969. *The Biology of Parasitic Flowering Plants*. Berkeley, CA: University of California Press.

Mower, J. P., S. Stefanovic, G. J. Young, and J. D. Palmer 2004. Gene transfer from parasitic to host plants. *Nature* **432**: 165–6.

Syvanen, M. and C. I. Cado (eds) 2002. *Horizontal Gene Transfer*. London: Academic Press.

Won, H. and S. S. Renner 2003. Horizontal gene transfer from flowering plants to *Gnetum*. *Proc. Natl. Acad. Sci. USA* **100**: 10824–9.

### Tracking the AIDS virus

Hahn, B. H., G. M. Shaw, K. M. DeCock, and P. M. Sharp 2000. AIDS as a zoonosis: Science and public health implications. *Science* **287**: 607–14.

Jenkins, G. M., A. Rambaut, O. G. Pybus, and E. C. Holmes 2002. Rates of molecular evolution in RNA viruses: A quantitative phylogenetic analysis. *J. Molec. Evol.* **54**: 152–61.

Korber, B. and 8 others 2000. Timing the ancestor of the HIV-1 pandemic strains. *Science* **288**: 1789–96.

Lemey, P. and 5 others 2003. Tracing the origin and history of the HIV-2 epidemic. *Proc. Natl. Acad. Sci. USA* **100**: 6588–92.

Li, W.-H., M. Tanimura, and P. M. Sharp 1988. Rates and dates of divergence between AIDS virus nucleotide sequences. *Molec. Biol. Evol.* **5**: 313–30.

O'Brien, S. J. and J. J. Goedert 1996. HIV causes AIDS: Koch's postulates fulfilled. *Curr. Opin. Immunol.* **8**: 613–18.

Ou, C.-Y. and 17 others 1992. Molecular epidemiology of HIV transmission in a dental practice. *Science* **256**: 1165–71.

### Chapter 7

#### Afrotheria theory

de Jong, W. W., A. Zweers, and M. Goodman 1981. Relationships of aardvark to elephants, hyraxes and sea cows from α-crystallin sequences. *Nature* **292**: 538–40.

Eizirik, E., W. J. Murphy, and S. J. O'Brien 2001. Molecular dating and biogeography of the early placental mammal radiation. *J. Heredity* **92**: 212–19.

Hedges, S. B. 2001. Afrotheria: plate tectonics meets genomics. *Proc. Natl. Acad. Sci. USA* **98**: 1–2.

Macdonald, D. 1984. *The Encyclopedia of Mammals*. New York: Facts on File Publications.

Madsen, O. and 9 others. 2001. Parallel adaptive radiations in two major clades of placental mammals. *Nature* **409**: 610–14.

Murphy, W. J. and 5 others 2001. Molecular phylogenetics and the origins of placental mammals. *Nature* **409**: 614–18.

Springer, M. S. and 6 others 1997. Endemic African mammals shake the evolutionary tree. *Nature* **388**: 61–4.

van Dijk, M. A. M. and 5 others 2001. Protein sequence signatures support the African clade of mammals. *Proc. Natl. Acad. Sci. USA* **98**: 188–93.

Zack, S. P., T. A. Penkrot, J. I. Bloch, and K. D. Rose 2005. Affinities of 'hyposodontids' to elephant shrews and a Holarctic origin of Afrotheria. *Nature* **434**: 497–501.

### Aussie songbirds

Barker, F. K., A. Cibois, P. Schikler, J. Feinstein, and J. Cracraft 2004. Phylogeny and diversification of the largest avian radiation. *Proc. Natl. Acad. Sci. USA* **101**: 11040–5.

Edwards, S. V. and W. E. Boles 2002. Out of Gondwana: the origin of passerine birds. *Trends Ecol. Evol.* **17**: 347–9.

Ericson, P. G. P., U. S. Johansson, and T. J. Parsons 2000. Major divisions of oscines revealed by insertions in the nuclear gene *c-myc*: a novel gene in avian phylogenetics. *Auk* **117**: 1077–86.

Ericson, P. G. P. and 6 others 2002. A Gondwanan origin of passerine birds supported by DNA sequences of the endemic New Zealand wrens. *Proc. R. Soc. Lond.* B**269**: 235–41.

Lovette, I. J. and E. Bermingham 2002. *c-mos* variation in songbirds: Molecular evolution, phylogenetic implications, and comparisons with mitochondrial differentiation. *Molec. Biol. Evol.* **17**: 1569–77.

Sibley, G. C. 1991. Phylogeny and classification of birds from DNA comparisons. *Acta XX Congressus Internationalis Ornithologici* **1**: 111–26.

Sibley, C. G. and J. E. Ahlquist 1986. Reconstructing bird phylogeny by comparing DNA's. *Scient. Am.* **254**(2): 82–3.

  1990. *Phylogeny and Classification of Birds.* New Haven, CT: Yale University Press.

### Madagascar's chameleons

Biju, S. D. and F. Bossuyt 2003. New frog family from India reveals an ancient biogeographical link with the Seychelles. *Nature* **425**: 711–14.

Brown, J. H. and M. V. Lomolino 1998. *Biogeography* (2nd edn). Sunderland, MA: Sinauer.

Lourenco, W. R. (ed.) 1996. *Biogeography of Madagascar.* Paris: Orstom .

Nagy, Z. T., U. Joger, M. Wink, F. Glaw, and M. Vences 2003. Multiple colonization of Madagascar and Socotra by colubrid snakes: evidence from nuclear and mitochondrial gene phylogenies. *Proc. R. Soc. Lond.* B**270**: 2613–21.

Pough, F. H. and 5 others 1998. *Herpetology.* Upper Saddle River, NJ: Prentice-Hall.

Raxworthy, C. J., M. R. J. Forstner, and R. A. Nussbaum 2002. Chameleon radiation by oceanic dispersal. *Nature* **415**: 784–6.

Roos, C., J. Schmitz, and H. Zischler 2004. Primate jumping genes elucidate strepsirrhine phylogeny. *Proc. Natl. Acad. Sci. USA* **101**: 10650–4.

Vences, M. and 6 others 2003. Multiple overseas dispersal in amphibians. *Proc. R. Soc. Lond.* B**270**: 2435–42.


The evolutionary cradle of humanity

Avise, J. C. 2000. *Phylogeography: The History and Formation of Species.* Cambridge, MA: Harvard University Press.
Cann, R. L., M. Stoneking, and A. C. Wilson 1987. Mitochondrial DNA and human evolution. *Nature* **325**: 31–6.
Goldstein, D. B., A. R. Linares, L. L. Cavalli-Sforza, and M. W. Feldman 1995. Genetic absolute dating based on microsatellites and the origin of modern humans. *Proc. Natl. Acad. Sci. USA* **92**: 6723–7.
Hammer, M. F. 1995. A recent common ancestry for human Y chromosomes. *Nature* **378**: 376–8.
Ke, Y. and 22 others 2001. African origin of modern humans in East Asia: A tale of 12,000 Y chromosomes. *Science* **292**: 1151–3.
Lewin, R. 1993. *Human Evolution: An Illustrated Introduction* (3rd edn). Oxford: Blackwell Press.
Takahata, N., S.-H. Lee, and Y. Satta 2001. Testing multi-regionality of modern human origins. *Molec. Biol. Evol.* **18**: 172–83.


Coral conservation

Fukami, H. and 6 others 2004. Conventional taxonomy obscures deep divergence between Pacific and Atlantic corals. *Nature* **427**: 832–5.
Knowlton, N. 1993. Sibling species in the sea. *A. Rev. Ecol. Syst.* **24**: 189–216.
Mace, G. M., J. L. Gittleman, and A. Purvis 2003. Preserving the tree of life. *Science* **300**: 1707–9.
Marcotte, B. M. 1984. Behaviourally defined ecological resources and speciation in *Tisbe* (Copepoda: Harpacticoida). *J. Crust. Biol.* **4**: 404–16.
Roberts, C. M. 2002. Marine biodiversity hotspots and conservation priorities for tropical reefs. *Science* **295**: 1280–4.
Veron, J. E. N. 2000. *Corals of the World.* Townsville, Australia: Australian Institute of Marine Science.


Sri Lanka, a cryptic biodiversity hotspot

Bossuyt, F. and 13 others 2004. Local endemism within the Western Ghats-Sri Lanka biodiversity hotspot. *Science* **306**: 479–81.
Myers, N., R. A. Mittermeier, C. G. Mittermeier, G. A. B. da Fonseca, and J. Kent 2000. Biodiversity hotspots for conservation priorities. *Nature* **403**: 853–8.
Somasekaram, T. (ed.) 1997. *Atlas of Sri Lanka.* Dehiwela, Sri Lanka: Arjuna Consulting.

Overseas plant dispersal

Baldwin, B. G., D. W. Kyhos, J. Dvorak, and G. D. Carr 1991. Chloroplast DNA evidence for a North American origin of the Hawaiian silversword alliance (Asteraceae). *Proc. Natl. Acad. Sci. USA* **88**: 1840–3.

Beverly, S. M. and A. C. Wilson 1985. Ancient origin for Hawaiian Drosophilinae inferred from protein comparisons. *Proc. Natl. Acad. Sci. USA* **82**: 4753–7.

Givnish, T. J. and 5 others 1996. The Hawaiian lobelioides are monophyletic and underwent a rapid initial radiation roughly 15 million years ago. *Am. J. Bot.* **83**: 159 [abstract].

Howarth, D. G., M. H. G. Gustafsson, D. A. Baum, and T. J. Motley 2003. Phylogenetics of the genus *Scaevola* (Goodeniaceae): Implications for dispersal patterns across the Pacific Basin and colonization of the Hawaiian Islands. *Am. J. Bot.* **90**: 915–23.

Tarr, C. L. and R. C. Fleischer 1995. Evolutionary relationships of the Hawaiian honeycreepers (Aves, Drepanidinae). In: *Hawaiian Biogeography*, W. L. Wagner and V. A. Funks (eds), pp. 147–59. Washington, DC: Smithsonian Institution Press.

Phylogenetic bearings on Polar Bears

Avise, J. C. 2005. Phylogenetic units and currencies above and below the species level. In: *Phylogeny and Conservation*, A. Purvis, T. Brooks, and J. Gittleman (eds), pp. 76–100. Cambridge: Cambridge University Press.

Cronin, M. A., S. C. Amstrup, G. W. Garner, and E. R. Vyse 1991. Interspecific and intraspecific mitochondrial DNA variation in North American bears (*Ursus*). *Can. J. Zool.* **69**: 2985–92.

Leonard, J. A., R. K. Wayne, and A. Cooper 2000. Population genetics of Ice Age brown bears. *Proc. Natl. Acad. Sci. USA* **97**: 1651–4.

Matsuhashi, R., R. Masuda, T. Mano, K. Murata, and Z. Aiurzaniin 2001. Phylogenetic relationships among worldwide populations of the brown bear *Ursus arctos*. *Zool. Sci.* **18**: 1137–43.

Paetkau, D., G. F. Shields, and C. Strobeck 1998. Gene flow between insular, coastal, and interior populations of brown bears in Alaska. *Molec. Ecol.* **7**: 1283–92.

Paetkau, D. and 10 others 1999. Genetic structure of the world's polar bear populations. *Molec. Ecol.* **8**: 1571–84.

Shields, G. F. and 8 others 2000. Phylogeography of mitochondrial DNA variation in brown bears and polar bears. *Molec. Phylogen. Evol.* **15**: 319–26.

Taberlet, P. and J. Bouvet 1994. Mitochondrial DNA polymorphism, phylogeography, and conservation genetics of the brown bear *Ursus arctos* in Europe. *Proc. R. Soc. Lond.* B**255**: 195–200.

Talbot, S. L. and G. F. Shields 1996. Phylogeography of brown bears (*Ursus arctos*) of Alaska and paraphyly within the Ursidae. *Molec. Phylogen. Evol.* **5**: 477–94.

Waits, L. P., S. L. Talbot, R. H. Ward, and G. F. Shields 1998. Mitochondrial DNA phylogeography of the North American brown bear and implications for conservation. *Conserv. Biol.* **12**: 408–17.

### Looking over overlooked elephants

Avise, J. C. 2000. *Phylogeography: The History and Formation of Species.* Cambridge, MA: Harvard University Press.

Comstock, K. E. and 6 others 2002. Patterns of molecular genetic variation among African elephant populations. *Molec. Ecol.* **11**: 2489–98.

Eggert, L. S., C. A. Rasner, and D. S. Woodruff 2002. The evolution and phylogeography of the African elephant inferred from mitochondrial DNA sequence and nuclear microsatellite markers. *Proc. R. Soc. Lond.* B**269**: 1993–2006.

Fernando, P. and 9 others 2003. DNA analysis indicates that Asian elephants are native to Borneo and are therefore a high priority for conservation. *PloS Biol.* **1**: 110–15.

Fleischer, D. J. C., E. A. Perry, K. Muralidharan, E. E. Stevens, and C. M. Wemmer 2001. Phylogeography of the Asian elephant (*Elaphus maximus*) based on mitochondrial DNA. *Evolution* **55**: 1882–92.

Roca, A. L., N. Georgiadis, J. Pecon-Slattery, and S. J. O'Brien 2001. Genetic evidence for two species of elephant in Africa. *Science* **293**: 1473–7.

### Bergmann's rule

Ashton, K. G. 2002. Do amphibians follow Bergmann's rule? *Can. J. Zool.* **80**: 708–16.

Ashton, K. G. and C. R. Feldman 2003. Bergmann's rule in non-avian reptiles: turtles follow it, lizards and snakes reverse it. *Evolution* **57**: 1151–63.

Ashton, K. G., M. C. Tracy, and A. de Queiroz 2000. Is Bergmann's rule valid for mammals? *Am. Nat.* **156**: 390–415.

Bergmann, C. 1847. Über die Verhältnisse der Warmeökonomie der Thiere zu ihrer Grosse. *Göttinger Studien* **1**: 595–708.

James, F. C. 1970. Geographic size variation in birds and its relationship to climate. *Ecology* **51**: 365–90.

Mayr, E. 1963. *Animal Species and Evolution.* Cambridge, MA: Harvard University Press.

Meiri, S. and T. Dayan 2003. On the validity of Bergmann's rule. *J. Biogeogr.* **30**: 331–51.

Queiroz, A. de and K. G. Ashton. 2004. The phylogeny of a species-level tendency: species heritability and possible deep origins of Bergmann's rule in tetrapods. *Evolution* **58**: 1674–84.

## Appendix

Avise, J. C. 2004. *Molecular Markers, Natural History, and Evolution* (2nd edn). Sunderland, MA: Sinauer.

Brooks, D. R. and D. A. McLennan 1991. *Phylogeny, Ecology, and Behavior*. Chicago, IL: University of Chicago Press.

2002. *The Nature of Diversity*. Chicago, IL: University of Chicago Press.

Cunningham, C. W., K. E. Omland, and T. D. Oakley 1998. Reconstructing ancestral character states: a critical reappraisal. *Trends Ecol. Evol.* **13**: 361–6.

Eggleton, P. and R. I. Vane-Wright (eds) 1994. *Phylogenetics and Ecology*. London: Academic Press.

Felsenstein, J. 1985. Phylogenies and the comparative method. *Am. Nat.* **125**: 1–15.

Fisher, D. O. and I. P. F. Owens. 2004. The comparative method in conservation biology. *Trends Ecol. Evol.* **19**: 391–8.

Freckleton, R. P., P. H. Harvey, and M. Pagel 2002. Phylogenetic analysis and comparative data: a test and review of evidence. *Am. Nat.* **160**: 712–26.

Garland, T., P. H. Harvey, and A. R. Ives 1992. Procedures for the analysis of comparative data using phylogenetically independent contrasts. *Syst. Biol.* **41**: 8–32.

Hall. B. G. 2004. *Phylogenetic Trees Made Easy: A How-To Manual* (2nd edn). Sunderland, MA: Sinauer.

Harvey, P. H., A. J. Leigh Brown, J. Maynard Smith, and S. Nee (eds) 1996. *New Uses for New Phylogenies*. Oxford: Oxford University Press.

Harvey, P. H. and M. D. Pagel 1991. *The Comparative Method in Evolutionary Biology*. Oxford: Oxford University Press.

Hennig, W. 1966. *Phylogenetic Systematics*. Chicago, IL: University of Illinois Press.

Huelsenbeck, J. P., R. Nielsen, and J. P. Bollback 2003. Stochastic mapping of morphological characters. *Syst. Biol.* **52**: 131–58.

Kolaczkowski, B. and J. W. Thornton 2004. Performance of maximum parsimony and likelihood phylogenetics when evolution is heterogeneous. *Nature* **431**: 980–4.

Maddison, D. R. and W. P. Maddison 2000. *MacClade 4: Analysis of Phylogeny and Character Evolution*. Sunderland, MA: Sinauer.

Martins, E. P. (ed.). 1996. *Phylogenies and the Comparative Method in Animal Behavior*. New York: Oxford University Press.

Martins, E. P. and T. F. Hansen 1997. Phylogenies and the comparative method: a general approach to incorporating phylogenetic information into the analysis of interspecific data. *Am. Nat.* **149**: 646–67.

Page, R. D. M. and E. C. Holmes 1998. *Molecular Evolution: A Phylogenetic Approach*. Oxford, MA: Blackwell.

Pagel, M. 1994. Detecting correlated evolution on phylogenies, a general method for the comparative analysis of discrete characters. *Proc. R. Soc. Lond* B**255**: 37–45.

1997. Inferring evolutionary processes from phylogenies. *Zool. Scr.* **26**: 331–48.

Price, T. 1997. Correlated evolution and independent contrasts. *Phil. Trans. R. Soc. Lond. B***352**: 519–29.

Purvis, A. and A. Rambaut 1995. Comparative analysis by independent contrasts (CAIC): an Apple Macintosh application for analyzing comparative data. *Computer Appl. Biosci.* **11**: 247–51.

Ricklefs, R. E. 1996. Phylogeny and ecology. *Trends Ecol. Evol.* **11**: 229–30.

Schluter, D., T. Price, A. Mooers, and D. Ludwig 1997. Likelihood of ancestor states in adaptive radiation. *Evolution* **51**: 1699–711.

# INDEX